T0408685

THE CUBE OF STRATEGIC MANAGEMENT

The Distinctive Advantage of Organizations

THE CUBE OF STRATEGIC MANAGEMENT

The Distinctive Advantage of Organizations

Mihai V. Putz

Apple Academic Press Inc.
3333 Mistwell Crescent
Oakville, ON L6L 0A2, Canada

Apple Academic Press Inc.
1265 Goldenrod Circle NE
Palm Bay, Florida 32905, USA

Exclusive worldwide distribution by CRC Press, a member of Taylor & Francis Group

International Standard Book Number-13: 978-1-77188-775-5 (Hardcover)
International Standard Book Number-13: 978-0-42939-764-6 (eBook)

Library and Archives Canada Cataloguing in Publication

Title: The cube of strategic management : the distinctive advantage of organizations / Mihai V. Putz.

Names: Putz, Mihai V., author.

Description: Includes bibliographical references and index.

Identifiers: Canadiana (print) 20190127813 | Canadiana (ebook) 20190127864 | ISBN 9781771887755 (hardcover) | ISBN 9780429397646 (ebook)

Subjects: LCSH: Strategic planning.

Classification: LCC HD30.28 .P88 2019 | DDC 658.4/012—dc23

Library of Congress Cataloging-in-Publication Data

Names: Putz, Mihai V., author.

Title: The cube of strategic management : the distinctive advantage of organizations / Mihai V. Putz.

Description: 1st edition. | Palm Bay, Florida : Apple Academic Press, 2019. | Includes bibliographical references and index.

Identifiers: LCCN 2019026275 (print) | LCCN 2019026276 (ebook) | ISBN 9781771887755 (hardback) | ISBN 9780429397646 (ebook)

Subjects: LCSH: Strategic planning. | Management--Technological innovations.

Classification: LCC HD30.28 .P88 2019 (print) | LCC HD30.28 (ebook) | DDC 658.4/038--dc23

LC record available at https://lccn.loc.gov/2019026275

LC ebook record available at https://lccn.loc.gov/2019026276

Apple Academic Press also publishes its books in a variety of electronic formats. Some content that appears in print may not be available in electronic format. For information about Apple Academic Press products, visit our website at **www.appleacademicpress.com** and the CRC Press website at **www.crcpress.com**

Dedication

To *Katy* and *Ella*:
The girls of my life!

About the Author

Mihai V. Putz, PhD, MBA, Dr.-Habil, is a laureate in physics (1997), with a postgraduation degree in spectroscopy (1999), and a PhD degree in chemistry (2002); with Post-Docs in chemistry (2002–2003) and in physics (2004, 2010, and 2011) at the University of Calabria, Italy, and Free University of Berlin, Germany, respectively. He is currently a Full Professor of theoretical and computational physical-chemistry at his alma mater, West University of Timisoara, Romania. He has made valuable contributions in computational, quantum, and physical chemistry through seminal works that appeared in many international journals. He is the Editor-in-Chief of the *International Journal of Chemical Modeling* (NOVA Science, Inc.) and the *New Frontiers in Chemistry* (West University of Timisoara). He is a member of many professional societies and has received several national and international awards from the Romanian National Authority of Scientific Research (2008), the German Academic Exchange Service DAAD (2000, 2004, and 2011), and the Center of International Cooperation of Free University Berlin (2010). He is the leader of the Laboratory of Computational and Structural Physical Chemistry for Nanosciences and QSAR at Biology-Chemistry Department of West University of Timisoara, Romania, where he conducts research in the fundamental and applicative fields of quantum physical-chemistry and QSAR. Among his numerous awards, in 2010, Mihai V. Putz was declared, through a national competition, the Best Researcher of Romania, while in 2013, he was recognized among the first Dr.-Habil in chemistry in Romania. In 2014, he was recognized by the Romanian Ministry of Research as Principal Investigator of the first degree at the National Institute for Electrochemistry and Condensed Matter (INCEMC), Timisoara, and was also granted full membership in the International Academy of Mathematical Chemistry. Recently, Mihai V. Putz expanded his interest to strategic management in general and to nanosciences and nanotechnology strategic management in particular; in this context, between 2015 and 2017, he attended and completed as the promotion leader an MBA on Strategic Management of Organizations—The Development of the Business Space specialization program at West University of Timişoara, the Faculty of Economics and Business Administration, while in between 2016 and 2019, he was engaged in the doctoral school of the same faculty, advancing new models of strategic management in the new economy based on frontier scientific inclusive ecological knowledge.

Contents

Abbreviations

8D	8-dimensional
BS	business space
CSR	competitive, sustainable, and regenerative
DD	democracy-to-democracy
DQD	double quantum dots
FDI	foreign direct investment
GCI	global competitiveness index
GDP	gross domestic product
IIA	international investment agreements
IPBES	intergovernmental platform for biodiversity and ecosystem services
NIRD	National Institute of Research and Development
QD	quantum dots
QKGR	quantitative knowledge–globalization relationship
QSAR	quantitative structure–activity relationships
R&D	research and developing index
RDI	research, development, and innovation
SDGs	sustainable development goals
SWOT	strengths–weaknesses–opportunities–threats
WIR	World Investment Report

Preface

In modern economy, management, and marketing, the *customer/client/ individual satisfaction,* coupled with organization income, comes first, no matter the environmental costs and social consequences. Then, currently, during postmodern times, while networks took advantage, social emerging rights and emancipation were continuously raised too, so social innovation takes preeminence, and the economy/management/marketing is thus re-oriented to comply with the needs, values, and *society satisfaction* as a whole—as a dynamic complex system, eventually featuring durability. However, only recently has serious consideration for the rare resources of environment, for example, in terms of renewable energy, species, weather, and even time, at both the global and local (GLocal) levels, appeared to be a serious issue. As a consequence, the proximity evolution (after post-modernism age, viz., "the future is here—yet not evenly distributed") in economy/management/marketing will be—highly probably—oriented to the *environmental satisfaction* as a whole, complex, inclusive, and sustainable GLocal system, instead.

In this context, the current top management, or the strategic management, has to shift so as to include the science and technology, not only as a support or adjuvant or object of business administration, but also to include it as an inherent part of the strategy building and of the Total Business Model. In other terms, the political economy (the management of welfare from/through organizations within and for nations) should be re-opened, re-defined to a bigger level, eventually to the ecological economy—to the benefit of holistic ecosystems. The ever-growing production and profitability goals of organizations should and probably will be replaced by recycling, renaissance, relaunching, remarketing, and reorganizing of any business, with the aim of the development of equilibrium to become the new main target, even starting far away from equilibrium or passing turbulent economic reality. The inclusive management and economy will nevertheless behave in a scientific manner; that means, it is evolving from its actual social science nature to the sum of interactions (the market) of individuals, producers, corporations, and nations; and all of which are re-projected into the economy driven by the interaction with the environment. Without exaggerating, one can step also toward forecasting the interaction with the cosmos too, namely, cosmoeconomy by the long-term planned Mars colonization, with aims of explorations—first, but then, passing to exploitation. However, the possibility of using solar energy from the universe, hydrogen from the universe, the gravitational force of the Moon, the Sun, etc., are currently implemented on a large social scale, either on the medium term, or shorter term from now. Such a new perspective requires new paradigms to be gradually introduced in both strategic management and business brand new planning.

This book is about to step in such direction. It advances the idea that eight-fold matrices (in 2D representations) and cubic realizations in 3D conceptualizations may offer a sufficiently complex model allowing in filling any conceptual market/space with a consistent degree of freedom, yet with symmetrical transformations and anticipative features. The eight-fold is a kind of magic algorithm, in strategy in general (think of a chess game, for instance, as an 8 × 8 strategic table) and also in management. Just remember the eight rules of management according to Peter Drucker's vision, that is, market standing, innovation, productivity, physical and financial resources, profitability, managerial performance and development, worker performance and attitude, and public responsibility. Equally, the eight-fold features make the search for excellence the true objective to be achieved by organizations, as projected by Peters and Waterman, needed is, a bias for action-active decision-making, being close to the customer/client, practicing autonomy and entrepreneurship, promoting productivity through people, practicing value-driven management, "sticking to the knitting," leaning the staff, and simultaneously having a loose-tight staff. Even needed personal development relating to efficient management, Stephen Covey identified firstly the seven good habits (i.e., proactivity, planning with the end/aim in mind, considering first things as coming first, implementing a win–win approach, understanding before being understood, synergizing—viz., competition, and refining all the personal and social capabilities—"sharpen the saw"), to which he then added the eighth step: the greatness (the inner voice inspiring others by leadership and personal significance model). The eight-fold paradigm crosses practically all intellectual fields, from the octave in music, to the octet in electronic occupancy in atomic elements, down to the chemistry periodic system, to the eight fundamental quarks, to the eight identified humankind dangers for the (post) modern age (i.e., the overpopulation—viz., competition, devastation of environment, man's race against himself, entropy of feeling—viz., moral hazard, the genetic decay, the break with tradition, indoctrinability, and nuclear weapons)—as identified by the Nobelist Konrad Lorenz for our evolving (ethological) civilization.

With such illustrious precedence, the present *research imparts* and advances the idea of *Cubic Strategic Management*: it basically consists of combining the triple point of competitive-sustainable and regenerative advantage into their +/− realizations, that is, the pros and cons as the driving force of the contemporary postmodern market: for example, features like competitive versus noncompetitive, sustainable versus nonsustainable, and regenerative versus nonregenerative. On the other hand, while passing from plane to space, on a 3D orthogonal referential, one deals with the self-producing of the cubic cell of eight strategies, at their turn identified according with the triplet of signs, in the simplified triple main actors of a business, that is, as the supplier, the producer, and the customer, which can be also rotated in their roles, respectively as (see also the figure below):

- (+1+1−1): Wise (including Ethical and Ecological) Business
- (−1+1+1): Underground (Parallel) Business
- (+1−1+1): Blue Ocean Business (Competitive Business)
- (+1+1−1): Smart (Win–Win–Lose) Business
- (−1−1+1): Perverted Spirits (Corrupted Business)
- (−1+1−1): Outsourcing (Lose–Win–Lose Business)
- (+1−1−1): Polemocracy (Win–Lose–Lose)
- (−1−1−1): Animal Spirits (Red Ocean or Lose–Lose–Lose Business)

Within the present cubic framework, essential and special management issues are approached (see the descriptions of the chapter below) and unfolded while new interesting features are found since reshaping many of the classical strategic management techniques. For instance, Michael Porter's five-forces competition model for organizations, as well as the Porter's (political) competition diamond for clusters/nations, is enhanced with new features, determinants, and analytical predictions for various organizations/clusters/ecosystems' cubic path's evolutions. This is because the cubic symmetry allows, for instance, rotations, mirroring operations, inversions, and translations—all of these being appropriately reinterpreted in the light of brand new forsaken economy/business planning and management. The strategic management in a cube idea appears generous, being here for the first time advanced, and with a great potential in developing both special and general strategies. The cubic paradigm allows, *inter alia*, working with fractional or interstitial features, deforming the cubes, intercalating networks by adjacent cubes, formulating various min/max paths, as well as "dramatically" organizing changes through performing symmetrical operations centered on specific elements (vertices—for firms, edges—for alliances, and faces—for clusters). The research here is entirely original, and it is a unique opportunity to publish it as a single flowing *corpus* instead of fragmenting it into various chapters and articles.

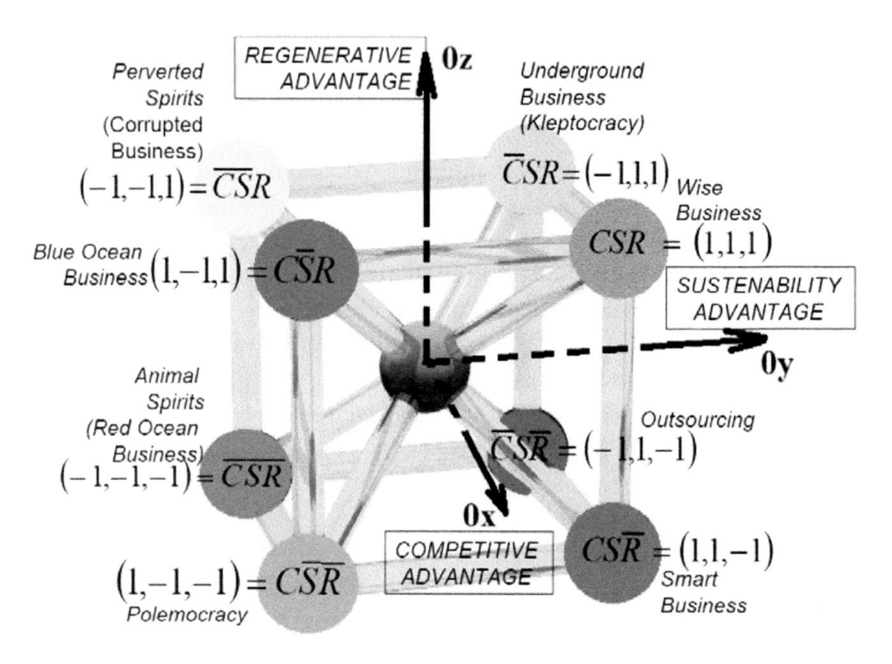

Nevertheless, the present research insight on cubic strategic management leads to the opening of a new area of research, say, the *geometrization of strategic management* with the "power/potential of the cube"; it also ensures the complex compactness as well as dispersion energies (viz., information vs. entropy in economy) of capabilities, resources, products, and services; it eventually allows the efficiency control and prediction (in terms of energy [kcal] and knowledge [bits]) by computation of each considered strategy as a min/max path within and between the cubes linked in a sustainable continuum (viz., the cubic chain value) of "coopetition"

(common values + competitive goods). It is an insight toward the holistic equilibrium (including the cyclic renaissance) for the current-future global society, with the belief that the coopetition between economy and environment can be foreseen, achieved, and endured!

—Mihai V. Putz
Timișoara, 2019

REFERENCES

Drucker, P. F. (1954, reissued ed. 2006) *The Practice of Management;* Harper Business: New York.

Peters, T. J.; Waterman, R. H. *In Search of Excellence - Lessons from America's Best-Run Companies;* HarperCollins Publishers: London, 1982.

Covey, S. R. *The Seven Habits of Highly Effective People,* FranklinCovey et co., Salt Lake City, 1989, 2004.

Lorentz, K. *Civilized Man's Eight Deadly Sins* (originally in German as: *Die acht Todsünden der zivilisierten Menschheit*), Piper Verlag GmbH: Muenchen, 1973, 1974.

Introduction

Management may be seen, nowadays, as the last resort in saving the world! The 21st century society last solution! This, because after the fail of ever-increasing industrialization, capitalization, society, economy, and ecology crisis, is a step to the corner! The diplomacy appears to be weakened by the competitive (often in a lose–lose game) strategy; the economy and international business appear to be driven by finance empowerment and gradual poverty as a counter-reaction; the military, "force keeping the peace," appears to use the technology against national natural developments; some regions clusterize; others go in fringes; the chaos is at the horizon, and the equilibrium is far away from the future present! Paradoxically, the only discipline, transdisciplinary, able to integrate all these dispersive tendencies, contrasting forces, and complex-often-chaotic reality is a management! Management may be seen as the new practical philosophy: the sociology without management is just doctrine (propaganda), while the management without the ecology (from individual to groups to nations) is just an illusion—susceptible to be dismantled at the first next disequilibrium! So, integrative management is the way, the new management, the scientific management built from algorithms that more and more drives our society: from production, to design, to artificial intelligence, life, comfort, and eternity! As a consequence, no effort is enough to imagine new management algorithm(s) able to integrate as much as of present life and being able to emulate the future challenges such as: to replicate itself and to multiply, to be close and open, to be local and global, central, and peripheral, simple and complex, punctual and circular, and so on, in dual opposites (and synergetic forces). As a whole, one may welcome the META-ECONOMY field as the *thesis-antithesis* [post-economy] *synthesis* in a Hegelian type construct; all of these for "bringing to society" the meta-economy in the same way as the triple perfect (trifecta) point mind–body–soul once dominated *meta-physics*! The search for triple perfect management (to be competitive in the business branch, to be sustainable to the environment, and to be regenerative in time) is the main goal of the new postmodern management project, to which the present book is also dedicated.

The book presents a complex and spanning fundamental, theoretical, and advancing new management model, responding to the current and to further 21st synergic interconnection needs in addressing the main societal, and humankind challenges, therefore, the information and communication, the renewable resources and energies, and the better and longer life (welfare).

Academia and open market researchers in management and marketing postmodern strategies are the main intended audiences; however, the book may be used also to:

- complement advanced university courses in environmental sciences and engineering (for Master and PhD courses) by addressing the frontier subject of management of sustainable strategies for the nanoscience and nanotechnology in the inclusive business model for a sustainable society and environment of 21st century;
- project a sustainable (inclusive) business model for scalable organizations (firms, companies, industries, clusters) while applying the compactness and variational principles in various markets (multidimensional space of values);
- model with the aid of the strategic cube of coopetitions (i.e., the cooperation in uphill and competition in downhill of the research-developing-innovation of a competitive product's life-cycles) the (post) modern economy/management of rare resources (energy, competence, and time), beyond the degenerative rivalry in meta-economy.

Accordingly, the main (indexing) keywords and concepts treated under the present advanced new management model refer to:

- Cross-Cultural Management
- Competitive Advantage of Organizations and Nations
- Strategic Cube of the Distinctive Advantage
- Extended Hamming Space
- Entropy and Information in Business Space/Marketing
- Maximum Path Principle in Technovation Management
- Contractors, Suppliers, and Logistics
- Game Theory and Postmodern Business Plan
- Industrial Districts and Clusters' Management
- Adjacency Matrix and Corporate Governance
- Composite Global Indices of Competitiveness (KOF, WIR, GCI, etc.)
- Change Management

They are, however, respectively, organized in 11 chapters, shortly described below:

Chapter 1 (*Cross-Cultural Management by Eight-fold Matrix of Social versus. Personal Values: Lessons from Generation-X*): The 8-dimensional space for the triple capabilities (triple perfect = trifecta points) are introduced. The context of cross-cultural management stands as an opening overture of the entire cubic theory in the book study; this way, the anthropologic-societal model for an organization, from Hofstede to Meyer, is considered toward the aim of the total management of quality. It provides the generalizing/reformulating possibilities of the individual-psychological profiles to the generational (sociological) ones; however, Generation X, for individuals and groups is taken as a case study with emphasis on the difference between organizational and societal responses on their actions thereof.

Chapter 2 (*The Strategic Cube of the Distinctive Advantage: An Epistemological Approach*): The competitive advantage is combined with sustainable advantage and with the regenerative advantage by the aid of the cubic algebraic transformations. This way, the reciprocal transformation and evolution are allowed due to the acquired cubic symmetries; moreover, one may unitarily comprehend the management of the Blue Ocean strategy in relation to the Red Ocean strategy, while re-inventing the strengths–weaknesses–opportunities–threats (SWOT) model and generalizing the generic Porter's forces of strategic management.

Chapter 3 (*The Strategic Cube of the Distinctive Advantage: Networks with Catastrophic Surfaces*): One takes the postmodern way of management, as based on the strategic cube developing distinctive advantage; the "concerted" network is orchestrated in three-dimensional space of competition. At this point, one further explores the cubic topological potentials while superposing them with the wave forms of (Thom's) universal catastrophes.

Chapter 4 (*Strategic Innovating Paths for the Distinctive Advantage: The Changing Management Faraway from Equilibrium*): One uses the strategic cube of distinctive advantage to make it applicable with the aid of the extended Hamming (-Putz) space. In such abstract space, the step-transformation or the econo-bits contribute to the maximum path or to the minimum entropy of the distinctive advantage, with special focus on innovation. This procedure allows for searching and enhancing of the negentropical hierarchy (also as a chain of *econo-wave-transforming behavior*, i.e., from "animal spirits" to "wise business") within the post-modern (far way from equilibrium), strategic, and meta-management.

Chapter 5 (*Scientific Entrepreneurship by the Strategic Double Cube of Competitiveness: Knowledge Transfer*): One advocates on how science and technology may be reconciled in the light of the strategic management, with the aid of the strategic cubic management model. As such, the scientific entrepreneurship models combine with the knowledge-transfer activities, toward complementing the "sacral" feature of science and "profane" use of technology in a new [meta-]economy and open society.

Chapter 6 (*Business Strategies by the Multi-nodal Logistics Within the Cubic Network of Distinctive Advantage*): Strategic management is employed by the hierarchy of a decisional tree, by which the vicinity culture and the networking logic are integrated into a strategic cube. This methodology builds a superior inclusive reality and post[/meta]-economy paradigm; in this context, the coupled uni-, bi-, and multi-nodal strategies may correspond with the cubic junctions, for instance, by their reciprocal vertices, edges, and faces, respectively; accordingly, the strategic thinking is this way enriched with new algorithms of the strategic cubic couplings.

Chapter 7 (*Risks Management in Nanotechnology Projects Toward Eight-fold Ws*): Risk analysis is specialized on the risk management level by combining the project management with a nanotechnological application. One may thus emphasize competitive, sustainable, and regenerative advantage by the eight-fold question marks (Ws) of a generalized and inclusive cubic strategic management thinking.

Chapter 8 (*Clustering in and out Strategies of the Prisoner Dilemma in the Cube of Distinctive Advantage*): One may be grounded on the (Italian) industrial district to introduce the dynamic clusters as a postmodern form of networking. Once the main players of a dynamical cluster as the (central) contractor and (chain) supplier are identified, game theory may be applied to decide on the prisoner's dilemma strategies. The resulting strategies widely integrate in the strategic cube of the distinctive advantage, while being qualitatively described by the adjacency matrix and quantitatively modulated by the information in the business space (and finally with the quantum attributes, in the sense of wave transforming).

Chapter 9 (*Strategic Innovation in the Organization Governance: The Eight-folding of the Mission Balance*): One advocates on the way in which the organization's governance should change its strategic mission, such that to be merely oriented to the third sector (the social services, including post-societal, including the environmental ones). As a result, the sustainability and regeneration are enhanced by post[meta]-economical [f]acts and products. At the same time, the informational energy becomes the strategic asset in the cube of the distinctive advantage, rather than the fashioned capital, through its localization/delocalization investments' dynamics.

Chapter 10 (*Global Strategies in the Knowledge Economy: The Case of R & D Sustainability in the European Union*): The ecopolitical–social index productivity index in the sociopolitic context, along with the investments index the world investment report index and competitiveness index global competitiveness index, are presented and described as suitable indicators in modeling the *knowledge–globalization relationships* (KGR) nowadays. However, on the quantitative level, the knowledge-global relationships appear as multilinear models so evolving from 1D to 2D to 3D orthogonally spaces. The output stands as the so-called Spectral-KGR deployment: it follows the principle of maximum path–minimum entropy in developing the organizations and business space within the strategic cube of the distinctive advantage.

Chapter 11 (*Cubic Management of Inclusive Scientific Change*): Change management is here modeled by identifying the axes of the change management: what to change? for the competitive axis, how to change?; for the sustainability axis, and when to change?; for the regenerative axis, respectively. Accordingly, the change management treats the change as an opportunity inscribed in the cubic management: one identifies eight successive steps in organizational, business, or general economical project deployment. Remarkably, the research–development–innovation organization is the first beneficiary of such construct; this is because they exploit the highest potential of changing. At this point, the fashioned *hierarchical model* of organization, while being further transformed into the Kotter's *dual model* of organization, may be turned into the present ultimate cubic model of organization, and *3D-change management* thereof.

However, the main leap and lesson of this monograph will hopefully bring a new perspective in treating the interconnected intra- and extraorganization, from micro to macro management, from local to global, and from center to periphery.

All of these, since, the cube, after all, may be seen in a postmodern eight-fold faces, multiplied by three capabilities on each, further connected, synergistically, and beyond, toward renewable, regenerative business and way of life, way of living, and way of future:

The Cube encloses and opens the space, the movement, the order!
The Cube inscribes and describes the evolution by paradigm,
gives the distinctive advantage, and sustainable progress!
The Cube is proactive: full and amply alike, compact,
and disruptive, 3D and projective!
The Cube is strategic: bureaucracy by repetitiveness,
democratic by symmetry, and meritocratic by selecting its path!
The Cube is pragmatic (economic), it is scientific (universal),
and it is artistic (social)!
The Cube has no beginning and ending (it is undulatory),
it is big and small alike (it is scalable),
and it is dissipative-conservative (it is applicative)!
The Cube is analogical (reproductive) and digital (unique),
it is natural and artificial, it is singular and multiple!
The Cube has no center and no periphery!

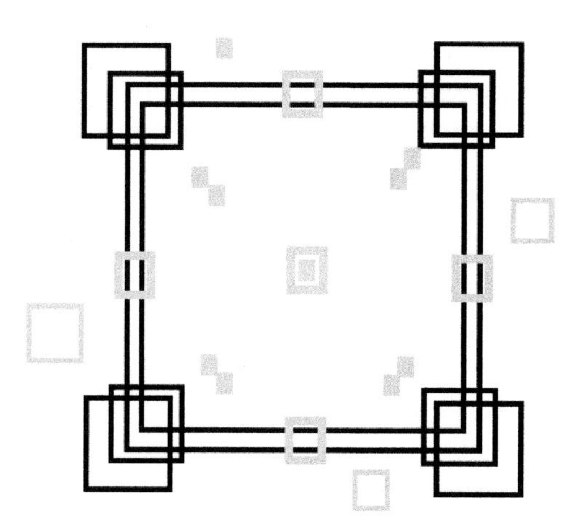

At the last but not at least important, the author thanks the Apple Academic Press team for embracing the project from its incipient phase through all revisions and in-house processing toward a top publication product, susceptible to becoming a referential one (and with possible forthcoming new editions and/or developments) in the new-management international literature of the 21st century! Long life to eco-ecolo-systems by GLocal-sustainability!

—**Mihai V. Putz**
Timişoara, 2019

Acknowledgments

To Professor Ecaterina Putz, PhD, for the classical and rational concerns in econo-mathematical and econo-physics calling the econo-chemistry frameworks!

To Professor Ioan Petrisor, PhD, for the inspirational debates "On the Heights of Despair" in the universal undulation of conceptual, trans-economic, and meta-management values!

To the National Institute of Research and Development for Electrochemistry and Condensed Matter—Timișoara (INCEMC—Timisoara, Romania), and to the developing National Research grant PN- PN-18-36-02-01/2018.

To the Executive Agency for Higher Education, Research, Development and Innovation Funding (UEFISCDI—Romania) for supporting the developing National Research grant PED 123/2017

Postface: An Imaginary Dialogue on Cubic Strategic Management

ON PRACTICAL MANAGERIAL SIGNIFICANCE

The Traditionalist: The cubic theory of strategic management assumes a very deductive approach and has applied an approach that has little empirical evidence to support. The literature review, of what exists, is very focused on establishing the credibility of the "cube" approach and the assumptions do not appear to be derived from the literature; it is rather a philosophical view held by the author. The kernel of the cubic theory has a place in the literature; perhaps, as an alternative to traditional approaches and for courses that are looking for unifying theories, it may have a place in higher education curricula.

The Progressist: The cubic strategic management is a cognitive research, a philosophical challenge on management, not induced by practical data, but deduced from the universal principles. Therefore, it should find its own place not only in postmodern management research and literature, but also in the higher educational level as such (from PhD probation and above). In any case, the time and further applicative research in strategic management will show where and when the "ethereal" contents of the book will find the "terrestrial" concepts of management.

ON INFORMATION TOPIC

The Traditionalist: There are competing rhetorical viewpoints that have tried to apply a unifying theoretical perspective to the social–technical systems and the complexities, which might add to complexity and chaos theories. The author basically assumes a preferred/existing approach and through conceptual/subjective analyses, applies it in a very acentric way. By applying the rules of cubic symmetry, the authors suggest that a proper alignment can be obtained through realignment of global forces of competitiveness. Assuming that this symmetry has any validity, what would those forces be like and how will they be properly managed? The cubic model and symmetry is interesting, but there are few, if any, references and citations to confirm its validity to the variety of applications that the authors are attempting to apply this model to. It is a classic example of a dogma applied to forcing a fit with a theory that may be suspected.

The Progressist: The basic education in the physics, known to us from Einstein's quotes, states that: "when the theory is not confirmed by the experiment, change the experimental set-up!" So, again, despite the apparent disconnection between the traditionalist (inductive) and progressist (deductive) approach of what should be the postmodern management, all "classical" theories of strategic management should be moved to face the new challenges and types of competitiveness existing in the 21st century. The cubic theory therefore augments the dynamic strategic management, while the cubic symmetries allow a rationale in aligning or moving the forces of competiveness. The cubic management is therefore oriented to what is called the strategic intent, rather than to classical strategic dynamics (which includes the diagnosis, scenarios, and deployment of a particular strategy). Nowadays, the strategies are diversified, combining, superposing, simulating, and dissimulating all competitive forces—once clearly distinguished (see the post-truth concept, so the post-strategy intent, and so on). The cubic theory has therefore the potential of opening a new way of thinking and understanding the strategic management, through symmetries and their dynamics, for which the "cube" is the first rational benchmark. On a later phase, it can, itself, be de-structured into various polyhedra, to the fractal seizure, the touch of the complexity systems, and social behavior being contained, in this way; but this would require a separate study. By the way, the "acentric" feature is also a very postmodern approach, which the reviewer maybe, nondeliberately, recognized to the book itself. Indeed, the postmodern age gives more weight "to the periphery than to the center," since the periphery dynamics may influence the "center of gravity," changing the entire strategy from periphery to the center (also, in relation with the chaos theory—equally asserting that a small modification, far away from the equilibrium center, may, eventually, change the systems' equilibrium)!

ON ANALYTICAL ASSESEMENT

The Traditionalist: Even deep machine learning principles assume the underlying principles of a system once they are discovered, and do not force an unrelated or proved approach on the data relational elements.

The Progressist: The "deep machine" is still a machine, so its learning is set algorithmically, following the analysis sequencing which, in methodological terms, is called a "thesis" that may nevertheless encounter (in practical issues) the "anti-thesis" with often cancelling effects, obscuring the true dynamics of learning. Instead, the present cubic management approach starts from the "synthesis" (in Hegelian sense, again recall that the traditionalist himself recognized the philosophical approach the present cubic management has undertaken), which can be manifested, either as a thesis or as an anti-thesis, but keeping the strategic dynamical intent intact, since coming from "above" (as the true strategy should be)—deductive, so proactive (otherwise, just reductive to the action and reaction, so reactive)!

ON HIGH-QUALITY SOURCE FOR OTHER RESEARCHES

The Traditionalist: Does the cubic theory assume that the unifying and simplistic symmetrical relationships exist, universally, without explanation?

The Progressist: One cannot agree that the symmetrical relationships in a cube are "simplistic;" remember, the cube is the next highly symmetrical architecture after sphere! And from the cube, all, and any other polyhedra, can be obtained by appropriate cuttings on vertices and edges, following its symmetries. Besides, making now a recourse to physical chemistry on how reality was modeled by deductive models, the atomic structure was initially assumed as the "cubic atom" by Gilbert Newton Lewis, before being consecrated as "spherical" by Niels Bohr—and even so, the Octet rules (maximum eight electrons in the corners of a cube) remained and provided a worked basis for what was later introduced as a chemical bond (modeled by linking vertices, edges, and faces of adjacent cubes, etc.). The same with strategic management—a move toward its true postmodern approach (in fact adapting to the postmodern and posteconomy age of these days) is needed. The cubic perspective was therefore a paved way which worked in the golden age of Physics in the 20th century! Why not challenge the strategic management of the 21st century with the same perspective!? At the end, the cubic paradigm has already been confirmed to have not only an universal but also an applicative impact on nature—it should have its part in the social sciences (and economy) too! The cubic theory is stepping toward this direction.

ON OVEREMPHASIS VERSUS UNDER EMPHASIS

The Traditionalist: There is an overemphasis that the model developed is a unifying theory and that its symmetry and its matrix algebra rules are universal. If you lose creditability in this, the entire theory falls short. There is an inherent underemphasis on the theoretical approaches that are the main stream in the applications discussed around each chapter.

The Progressist: This is the way a cognitively based (or philosophically based, as also the traditionalist earlier recognized) approach is undertaken. They appeal to a benchmark, and a universal one, the "cube," in this instance; it will be hard to deny the "power of the cube" in centralizing, synthesizing, analyzing, and destructing the contents it circumvents—since this was already proved by the entire scientific background and history (viz., the overemphasis)—with on-life applicative impact too (viz., the underemphasis), see the "cult-like" Preface, too. Now it is the turn of strategic management to take this road—the cubic theory stimulates toward this direction, as a research, as a promise, and as a strategic intent approach.

ON CREDIBILITY

The Traditionalist: Does the prioritizing of weights and relationships demonstrate matrix *algebra compelling arguments?*

The Progressist: The cubic theory aims to push forward the academic community, and does not try to convince, by finding specific numbers to sustain its universal approach, which in any case will be limiting (i.e., we will never be able to prove by applications that a theory is valid in an universal way, i.e., in all present and future cases); so, the safely science recourse is to advance paradigms, as does this one, too.

ON OPPORTUNITY

The Traditionalist: The cubic theory appears to be opportunistic!

The Progressist: This is not necessarily bad—the strategy itself should imply that even in the classical SWOT analysis (what [O]pportunities I do recognize that using the company [S]trengths to diminish the company [W]eaknesses, so that the [T]hreats be avoided?). Thus, the "opportunistic" feature is in the DNA of strategic management too; it is its key, and this cubic approach may use it to provide the new strategical intent. It is also speculative, but this is again well aligned with the generic economical behavior.

ON SUSTENABILITY EDUCATION

The Traditionalist: There must be buy-in from the academic community that the cube symmetry has a merit in.

The Progressist: I do believe, also based on the previous traditionalist comment, that is, the theory may be seen "as an alternative to traditional approaches and courses that are looking for unifying theories, it may have a place in higher education curricula." Indeed, such an approach was missed so far in the community market; so, it definitely will fill and hopefully open a new perspective of strategic management, perhaps deductive—perhaps approaching the way the "global management" will turn the "posteconomical" competitiveness in a more aligned way, as also the traditionalist earlier recognized (i.e., by the "realignment of global forces of competitiveness") in a truly inspiring way!

ON GENERATION-X

The Traditionalist: The cubic theory has the most unorthodox introduction to the topic one can imagine: the concepts of modern management of intercultural communication and the generation-X were interesting, but apparently extremely disjointed.

The Progressist: Perhaps it is not just a chance that the generation-X was born with the postmodern management, while arriving now at joint maturity: what seemed to be disjoint at beginning of their "life cycle" becomes inter-related at the maturity. Nevertheless, with the help of cubic management, seeded in the eight dimensions of generation-X too, it may be regarded as a leverage to the second curve, re-launching management too, to the next generations, frontiers, and beyond.

ON STRATEGIC MANAGEMENT INTENT

The Traditionalist: The cubic theory may seem static, more related to the planning function of management, and thus appearing somehow reductive for the strategically management intent. Does it have the inner dynamical nature? Can this be "saved?"

The Progressist: Once Plato replayed to some of his opponents that "you can see the horse-animal, yet you must have the special eye to see the horse-idea!" The same here: the cube may only seem static since it has high symmetry. In fact, once it is "populated" with "colors, entities, and behaviors" on its corners, they develop specific proximity fields, which at their turn interfere, constructively and destructively, giving a whole dynamics (life) to the cubic space itself! Even more, think of the NaCl crystal, with its ionic sub-lattices; they are universally in interaction to maintain the perfect cube in an ideal condition (no perturbation). Once placed in specific open conditions, the perfection for the cube is deformed under the external conditions (pressure, temperature, for instance); still, their universal interaction is the same. So they will tend to always regain the initial perfect state, thus developing the inner tension "fighting" the open external conditions; so the evolution in a broad sense is always present! The same is also true with the cubic model of strategic management: it is populated by various behaviors, types of organizations, costs, resources, risks, and weights of economy. They all interact, according with the cubic symmetries, and develop the inner tension, a creative tension, to find their ideal reciprocally position such that they "all can survive" in perfect cubic equilibrium under potentially external, asymptotic, or by distance actions (a.k.a. economical crisis, turbulence, and chaos)! This way, the cubic strategic management naturally contains dynamics, strategically intent, and universality alike!

ON FURTHER REFERENCES

The Traditionalist: One can generally refer starting from the strategic management areas. However, the proposed cubic theory covers such a wide spectrum of topics that the proposed further references list is sparse indeed; here it is just an addendum by various topics:

ON STRATEGY AND SUSTAINABILITY

Hamel, G.; Prahalad, C. K. Strategy as Stretch and Leverage. *Harv. Bus. Rev.* **1993,** *71* (2), 75–84.

Michalisin, M. D.; Smith, R. D.; Kline, D. M. In Search of Strategic Assets. *Int. J. Org. Anal.* **1997,** *5* (4), 360–387.

Michalisin, M. D.; Smith, R. D.; Kline, D. M. Intangible Strategic Assets and Firm Performance: A Multi-Industry Study of the Resource-based View. *J. Bus. Strateg.* **2000,** *17* (2), 91–117.
Porter, M. E.; Kramer, M. R. Strategy and Society: The Link Between Competitive Advantage and Corporate Social Responsibility. *Harv. Bus. Rev.* **2006,** *84* (12), 78–92.
Scott, S. Corporate Social Responsibility and the Fetter of Profitability. *Soc. Responsib. J.* **2007,** *3* (4), 31–39.
Theriou, N.; Aggelidis, V.; Theriou, G. A Theoretical Framework Contrasting the Resource-based Perspective and the Knowledge-based View. *Eur. Res. Stud.* **2009,** *12* (3), 177–190.

ON GREEN AND ECO-FRIENDLY SUSTAINABILITY

Barney, J. Firm Resources and Sustained Competitive Advantage. *J. Manage.* **1991,** *17* (1), 99–120.
Carter, C. R.; Easton, P. L. Sustainable Supply Chain Management: Evolution and Future Directions. *Int. J. Phys. Distrib. Logist. Manage.* **2011,** *41* (1), 46–62.
Cavaleri, S. A. Are Learning Organizations Pragmatic? *Learn. Org.* **2008,** *15* (6), 474–481.
Harrison, J. S.; Bosse, D. A.; Phillips, R. A. Managing for Stakeholders, Stakeholder Utility Functions, and Competitive Advantage. *Strateg. Manag. J.* **2010,** *31* (10), 58–74.
Husted, B. W.; Allen, D. B. *Corporate Social Strategy: Stakeholder Engagement and Competitive Advantage*; Cambridge University Press: Cambridge, UK, 2011.
Mitchell, R.; Agle, B.; Wood, D. Toward a Theory of Stakeholder Identification and Salience: Defining the Principle of Who and What Really Counts. *Acad. Manag. Rev.* **1997,** *22* (4), 853–886.
Porter, M.; Van Der Linde, C. Toward a New Conception of the Environment: Competitiveness Relationship. *J. Econ. Persp.* **1995,** *9* (4), 97–118.
Summers, G. J.;Scherpereel, C. M. Decision Making in Product Development: Are You Outside-in or Inside-out? *Manage. Decis.* **2008,** *46* (9), 1299–1314.
Wagner, M. Green Human Resource Benefits: Do They Matter as Determinants of Environmental Management System Implementation? *J. Bus. Ethics* **2013,** *114* (3), 443–456.
Zhu, Q.; Sarkis, J.; Lai, K. Green Supply Chain Management Implications for 'Closing the Loop.' *Transp. Res. Part E* **2008,** *44* (1), 1–18.

The Progressist: This way, the cubic theory may be referred to, or confronted with, as will be the case, the "classical" theories of strategic management—that, usually, is implying the diagnosis, and deployment, the optimization and efficiency; however, each time the "acentric" [ex-centric] perspective will have to be faced, a highly original and challenging potential, the symmetry will be needed, while the present cubic management will stand along that side to positively generalize, unify, and interconnect, with the intent to reopening the managerial custom reports, expectations, and beliefs.

CHAPTER 1

Cross-Cultural Management by Eight-Fold Matrix of Social Versus Personal Values: Lessons from Generation-X

ABSTRACT

The transcultural scenery on the society, organization, and at the individual level, may be unified, from the differences/associate dimensions perspective, on the eighth degree unfolding. By appropriate adapting of the anthropologic–societal model for an organization, from Hofstede to Meyer, one may formulate the aim of the total management of quality alongside with generalizing/reformulating the individual–psychological profile to the generational (sociological) one. Thus, the premises for a comparative analysis are offered, by means of a transcultural qualitative–phenomenological matrix, as a basis/guide for a quantitative substantiation (through scores based on a hypothetic interview). Also, for a correlated analysis, whenever possible, one may estimate the degree in which the cultural model is realistic, while choosing between the second culture manifestation level on personal–generational, professional and societal–national manifestation forms. The application on the generation-X is unfolded by the 8D-phenomenological correlations with the societal and organizational environment; the model can be extended on the other generations too, as well as, on the multigenerational aggregate system, once referred to as the societal, or organizational context. All the efforts should conduct to an appropriate multidimensional strategy for communication, promotion, employment within organization, public relationship, etc.

Motto:

"**Lt. Joshi**: *'The World is built in a wall that separates kind. Tell either side there's no wall, you've bought a war. Or a slaughter.'*"
—Blade Runner 2049 (2017) © Alcon Entertainment, Columbia, Free Scott, and Warner Bros. Pictures & Co.

1.1 INTRODUCTION

In the global and complex world at planetary level, individual–organization–society interaction is crucial, for both personal life and the well-being of the economy and society in general. In this context, psychology—at the individual level, with sociology—at the level of organizations, and anthropology—at the societal/nation level intertwines to various levels of significance and on various cultural dimensions (within the so-called "second culture"), creating a cross-cultural "landscape." Here, the term "transcultural" also signifies (Nicolescu, 2007; David, 2015):

- between-cultural levels (individual–organization–society)
- through-cultural levels (intralevels)
- beyond-cultural levels, in the sense of synergy between levels (including their correlation).

For these cultural levels, the current status is as follows (Fig. 1.1):

- The individual (psychological) level: It is manifested through the so-called OCEAN acronym (Mc Crane and John, 1992), openness to experience, conscientiousness, extroversion, agreeableness, neuroticism—also called the Big Five individual behavioral traits; to these, the sixth dimension emerged precisely from the participation of the individual, in organizations and society, as a kind of feedback from higher individual levels, which influence their behavior, attitude, and in a smaller degree even the character; it is due to Hofstede (2007) and refers to addiction to others.
- The social (anthropological) level: It manifests equally to every individual, inherent part of a society, or nation. With even stronger influences than the organizational ones (those related to the place of the profession), they are due to the profound elements related to birth, language, family education, myths, traditions, etc., being

much more powerful than organizational culture (temporary or easily transformable, changed, abandoned, etc.). At this level, Hofstede's works (on statistical basis) are the most influential, imposing four, five, and finally six cultural differences at the social or national level reflected in the appropriate cultural dimensions: the distance to power, individualism versus collectivism, masculinity versus femininity, avoidance of uncertainty, to which the long-term orientation versus short one was also added. The Confucianism and subsequent indulgence versus constraint associated with dependence on others value completes Hofstede's cultural landscape (Hofstede et al., 2010). However, these dimensions have recently been reformulated and extended to 8-dimensional (8D) perspective of cross-cultural management, which are as follows (Meyer, 2014):

- communication (in a high and low context, taken from the American anthropologist Edward Hall);
- assessment and criticism (constructive, frankness, or implicit-diplomatic approach);
- power of persuasion (through algorithms and specific-analytic, or holistic global synthesis arguments, e.g., European analytical thinking vs. Asian holistic thinking, etc.);
- leadership (respect and deference to authorities, correlates with distance to power);
- decision (consensus-oriented, according to the hierarchical organization, correlates with collectivism vs. individualism, e.g., the Germans decide hierarchically, the Americans in work groups; the Japanese are both strongly hierarchical and strongly consensual);
- trust and truth (in the governing bodies and the organization of the society as a whole, combining the emotional trust, "with the heart," with the rational one, "with the head" in the rules and traditions of society, with a tendency toward mutual trust);
- conflict and misunderstanding (the contradiction is healthy, the way to progress, the dubito ergo sum, the mechanism of social revolutions, the change of the "class struggle," respectively, the tolerance for divers and new, correlate with the avoidance of uncertainty);
- time programing (correlated with the monochronic/chromatic and polychronic/polychromatic time in Hall's classification,

respectively, with short- and/or long-term orientation, in Hofstede's dimensions).

- Organizational (sociological) level: It is the most "volatile" and in the change of the cultural level, being tributary to economic conceptions in a certain historical period (less than what it really defines, anthropologically, a society/nation). It is subjected to local and global changes and depends on the internal and external conjuncture of the organization (forces, resources, latent potentials, changing national and international political laws and systems, competition, technological and climate changes, etc.). Values and cultural differences/dimensions can be diverse, for example, income inequality, respect for the elderly, political violence, number of laws and directives, expert confidence, xenophobia, rapid leadership (vehicles and decisions), well-being, fast speaking (but also thinking and decision-making), close ties with the family and organization (seen as an extended family), the personal assumption of authority and truth (frequent use of the word "I"), the pressure for economic growth (and profit), low/high percentage of women in organizations (and in management), exchange rates, investing in poor areas, adaptation to changing reality, concern for social obligations (wise business), pride of the organization, isolationism, sports activity, diet at work, professional Internet, security at the organization level, etc. Although many of these cultural variables correlate among them, and with dimensions in the social or anthropological category, one may seek to the essential reduction of these values or cultural dimensions. To this aim, let us focus on the total quality management of an organization; accordingly, an organization (core) values' reduction can be performed and one may distinguish the 8D ones (Detert et al., 2000):
    - the basis of truth and rationality in the organization (management by facts, realities, scientific data, logical and strategic models, etc.);
    - nature of the time and the time horizon (very short-term operationalization; medium and long-term planning);
    - motivation (see Maslow's pyramid with adapted variants, adjusting errors, misunderstandings, and conflicts);
    - stability versus change/innovation/personal development (strategy through differentiation, diversification, risk taking,

education/continuous improvement, optimization of production cycles, etc.);
- orientation toward work goals–team workers (productivity, management through objectives, through projects, etc.);
- isolation versus collaboration and cooperation (promotion, alliances, subsidiaries, investment strategies, disinvestment, niche business, etc.);
- control, coordination, and responsibility (internal rules and procedures, product- vs. process-oriented management, common values and qualities of product and human resources, etc.);
- strategic orientation and focus on the inside/outside of the organization (human resources, diversity vs. homogeneity, competition forces—consumers, suppliers, substitute products, other public or private organizations, etc.).

Having been exposed to the three transcultural levels, with their dimensions, some fundamental problems can be solved. Especially, the cognitive and methodological analyses are relevant for the organizational strategy; the organization may be considered as "the middle of the distance" between the individual and the society or nation, yet, also, offering personal and public or national development, depending on the complexity of the qualitative response (usefulness) to society. The impact may be measured by means of phenomenological (qualitative) and statistical (quantitative) deployment. The present study provides the basis for such a perspective, respectively:

- In Cognitive Analysis section, the arguments of an 8D-transcultural unification are provided, with the listing of the eight dimensions for societal values and the generalization of individual ones, for generational identity, with generation-X as a case study.
- In Methodological Analysis section, the intercultural, societal–generational, and organizational–generational phenomenological matrices for generation-X are formed.

1.2 COGNITIVE ANALYSIS

There are some fundamental cognitive questions (and of knowledge in the broad sense) to provide arguments for any transcultural approach, and for the present one as well (Hofstede, 2011):

- Why are cultural dimensions necessary? Cultural dimensions are intangible, they are mental constructs; however, the differences they express in thinking, expression, and action justify the study of behavioral patterns at the level of individual, organization (groups of individuals), and society/nations (clusters of groups).
- Why more cultural dimensions are needed? Because individual, organizational, and societal stereotyping on a small number of dimensions can induce confusion, especially in managerial approaches, when cosmopolitan differences influence the competitive, sustainable, and regenerative advantage of the products and services offered, promoted, etc.
- How many cultural dimensions are needed? The debates are open: there is a trend of diminishing dimensions (supported by Hofstede), whose followers are nonetheless systematically "forced"/convinced to accept additional dimensions; thus, ideal classifications should be restricted to approximately seven categories of differences plus/minus two dimensions (Miller, 1956); Hofstede thinks that it is rather a "minus" dimension, whereas we think it is rather "+1 or +2" needed dimensions; this, also, is because applying only "−2" we reach five specific categories, in equal cardinal, to the dimensions of the individual's psychology. Instead, additional dimensions should be added when starting from the individual and treating groups (organizations), and then groups of groups (meta-groups), or masses of individuals (society or nation). However, Hofstede himself admitted the sixth dimension (the dependence on others) as being common to the individual and to the societal culture; the result is in accepting the need of "+" over the basic dimensions of the individual; we believe that this sense must be maintained in order to render the complexity (controlled or at least understood) of dynamic systems at the level of organizations and society.
- How the eight transcultural dimensions are used (useful)? Because it is unifying on the current dimensional characterization at the society level (Meyer, 2014) and on the management based on total quality in organizations (Detert et al., 2000); in addition, the 8D model can be rationalized into the cubic structure (the cube has eight corners), and from there to the structural networks that can represent and shape interactions (peaks, sides, or faces) between individuals, organizations, societies, but also transculturally, at the

individual–organization–society correlated levels; in addition, here we also meet Hofstede's demand for cultural diminution, the 8D model can be obtained from a "three-dimensional, 3D" nucleus called trifecta point (three times perfect, see Trinity) by aggregated combination of positive and negative expressions (corresponding to the positive and negative axes "•" in a given direction), thus, giving a point with the coordinates of the trifecta point (a, b, c), it also generates seven distinct points, namely: ($\underline{a, b, c}$), ($\underline{a}$, b, c), (a, $\underline{b}$, c), (a, b, $\underline{c}$), ($\underline{a}$, $\underline{b}$, c), ($\underline{a}$, b, $\underline{c}$), and ($\underline{a}$, $\underline{b}$, $\underline{c}$) for a total of eight points arranged in the corners of a "cultural cube"; hence, the transcultural, inter-/between-, and synergic/beyond the cultures of the individual, organization, and society appear in an algorithmic manner, of strategic, personal, economic, and public potential, respectively (Putz, 2019).

- How can the individual psychological level (intrinsically unary and individual monads) be correlated with the levels of organizational and societal culture (which are also sociological and anthropological representations of human collectives and groups)? Here, there is indeed the danger of correlating the values of individuals with group-specific cultures, a logical and methodological inconsistency that has long been reported in the literature (Robinson, 1950) under the name of "ecological fallacy"; nevertheless, such correlations have been made by "forcing the note" from a phenomenological point of view, as in the study of Hofstede and McCrae (2004); for example, for the 30 countries surveyed in a statistical study, 39% of extrovertism is explained by individualism, 31% of neuroticism is explained by avoidance of uncertainty, whereas 55% of neuroticism is explained by the combination of avoidance of uncertainty and masculinity. However, in order to answer the initial question, we prefer to generalize the individual values to the generational values, to which the individual belongs; on the other hand, it is true that the complementary and aggregated individual values "dissolve" and adapt, according to the generational group to which they belong; at the same time, the specific-group values are more prominent on generation than for the same type of invariant behavior, between generations (very unlikely). We generalize and practically "dissolve" the values of the psychology of the individual in the generational psychology to which he belongs, which he now represents him as a group, and thus sociologically compatible (as a nonunary number) and as behavior (anthropological, in the sense of collective psychology, generational here) with the possibility of cross-cultural correlations, avoiding "ecological error." This approach will be illustrated for generation-X (born between 1965 and 1980) in the next section.

1.3 METHODOLOGICAL ANALYSIS

In the modern management of the "golden" rule of intercultural communication: "what you do not like, do not do it to others!" is replaced by the so-called "titan" rule: "do to the other, according to his preferences!" (AARP, 2007).

We continue the cognitive analysis in the previous section, specifying the trifecta points for the three cultural second levels in Figure 1.1, but with the generational level replacing the individual level as in Figure 1.2. We therefore identify the transcultural trifecta components as shown in Table 1.1.

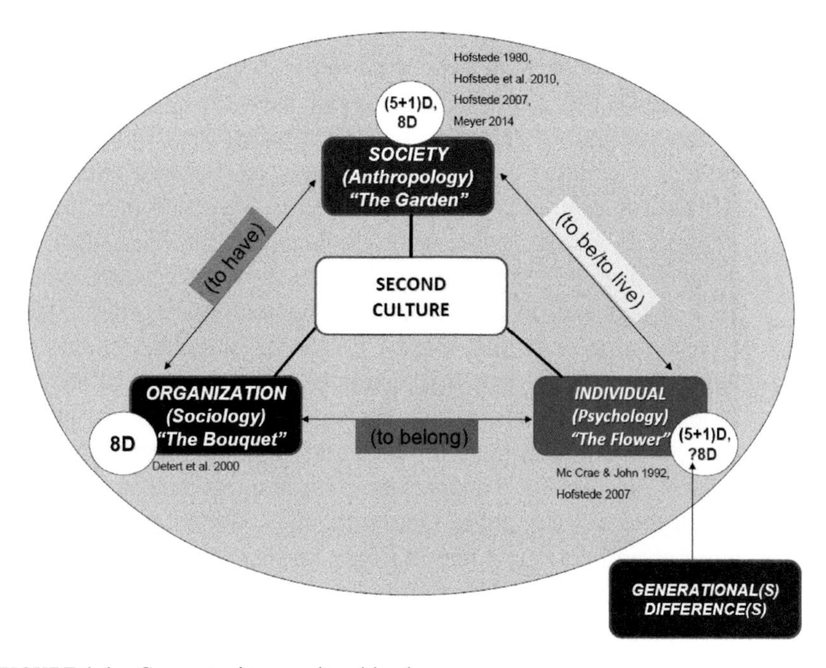

FIGURE 1.1 Conceptual transcultural landscape.

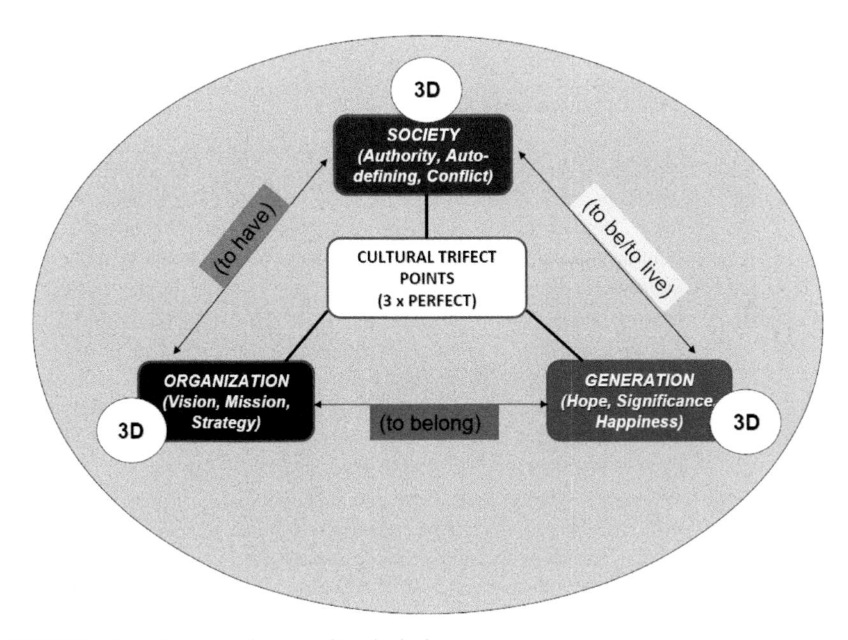

FIGURE 1.2 Conceptual transcultural triptic.

TABLE 1.1 Trifecta Points for the Levels of Generational, Organizational, and Societal Cultures.

Culture	Generational	Organization	Society
Coordinates of trifecta points	Hope	Vision	Authority
	Significance	Mission	Self-defining
	Happiness	Strategy	Conflict

Trifecta points are 8D generators, through combination on all the positive/negative space available for each cultural level. The "spectrum of values" for each cultural level ("cultural octave") in Figure 1.1 is shown in Figure 1.3. Note that generational values are customized and classified for generation-X in Table 1.2, based on the characteristics of this generation (Williams and Page, 2011).

From the corroboration of the characteristics of the cross-cultural dimensions at the level of the generation-X with the 8D-organizational and 8D-societal values, the qualitative correlations are shown in Figures 1.4 and 1.5. For the societal–generational X (S–GX) qualitative correlations in Figure 1.4, the following phenomenological features are distinguished:

TABLE 1.2 Dimensions of Generational Culture and Their Development for the Case of Generation-X.

Generational dimension	Manifestations in generation-X
Implication and family	Traditional family persons (grew up with a high rate of parental divorces, contemporary with AIDS/HIV), respect religious values, "what is a hero?" hard to impress, adorable on the phone, preferably at work, volunteer, proactive (outside the organization is often misinterpreted as nonloyalty)
Work and career	Higher education and independence; preference of tasks on time but "in their program," tendency for advanced positions and services, "work to live," careers are "portable," service is "just a job," work is "a challenge," but bring plus-value to the workplace, "multitaskers," generational bridges "between boomers and Y-chi, millenarians, etc., open to self-development," work with you—not for you
Recognition (financial included)	Pragmatic, ignoring the leaders; they are unimpressed by authority (those in the previous generation are considered "outdate" and sometimes corrupt, and those after them are considered to be "inappropriate and easily manipulated" or "moral hybrids"), recognize the merits of others and expect to have their merits recognized, an increased sense of "what they deserve," but not "in love" with public recognition
Freedom (loyalty included)	Distrust of politicians and leaders (contemporary with Watergate, the end of the Cold War), and with propensity for nonloyalty (following "betrayal"), but loyal to the manager (the direct collaborator who demonstrates skills and abilities) with a high critical sense, hungry for flexibility of their personal work schedule
Discipline and logics	Predominantly global thinking, know to keep "secrets," cautious, conservative, but fair and rational, organized, good managers, want/await feedback, but reject the imposed rules
Change and diversity	Have grown up with computer, energy crises, political changes, music changes (technomusic), respect diversity, they sometimes look for it, they know how to have a good time (sometimes at work, too), adaptable, willing to innovate, expect the latest information
Balance	Skeptical, cynical, but deep down in their souls (with a hope for "better"), but not as optimistic as Ys (those belonging to Generation-Y) after them, balance their personal and professional life (they are not workaholic as their "boomer" parents), self-control, self-confidence, positive approach to life
Motivation (focus included)	Entrepreneurship spirit, meritocracy, women want to work at a job and not at home, having time as value, focusing on goals, on project and results; they are expecting to be properly rewarded; eager to save money, time, and energy, likely to develop strategies to stay "marketable," manifesting versatility in personal and professional offers, approachable with humor (and "tricked"), if allowed to make a difference, accepting the sabbatical year

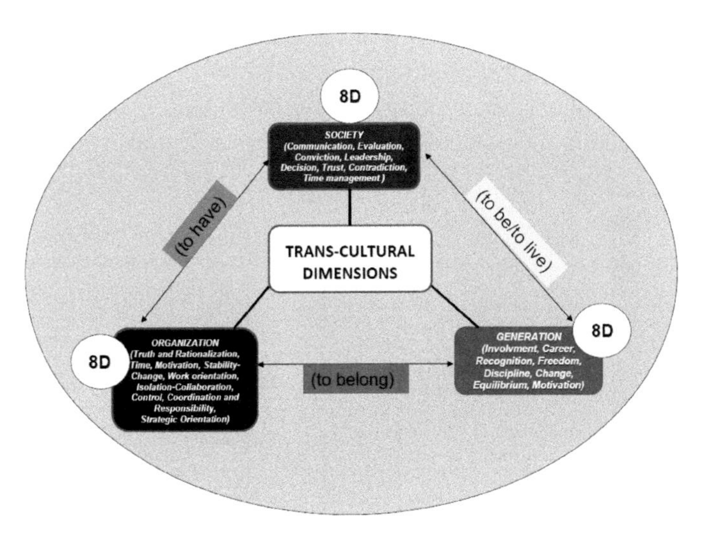

FIGURE 1.3 Conceptual transcultural 8D-universe.

- The values of generation-X correlate qualitatively in half-moon (parabolic) shape, either as strong ("full-colored targets"), or as weak (through "uncolored targets"); in addition, the strongest correlations are to be found for the couples: recognition–time, freedom–trust, change–contradiction; the least correlations are in the couples: involvement–decision, career–persuasion, discipline–assessment, balance–communication, and motivation–leadership.
- The organizational–generational X-qualitative correlations are represented in Figure 1.4, where the following phenomenological information is distinguished:
- Generation-X values correlate qualitatively in the shape of an arrow (linearly), with the most correlated G–SX couples (excluding the overlapping ones, of type motivation and change): freedom–time, discipline–strategy, and the least correlated are: involvement–control, career–isolation, recognition–work, balance–truth, and rationality.

These qualitative phenomenological correlations allow, in a later stage, the "coupling" in the cubic representation upon the "nodes, edges, or faces" of the cubes (bearing 8D information) for the societal/organizational/generational transcultural levels (X in the present case). Some further studies can also unfold the present issue in a more complex way, by considering structures (as multigenerational organizations, as they are in fact) extended into networks; the development of scenarios and, therefore, strategies and organizational (and societal–public policies, for example) are therefore algorithmic, on a scientific basis, that is, based on dynamic networks, within the postmodern sense of strategic management.

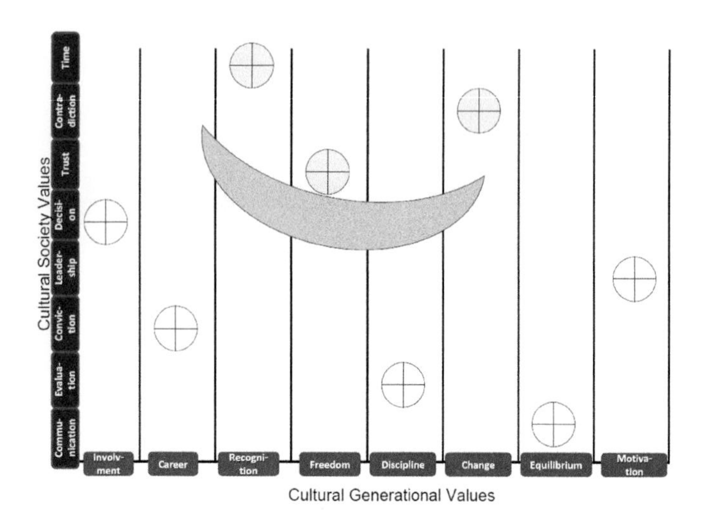

FIGURE 1.4 Intercultural matrix society versus generation-X.

In addition, the present study can be further substantiated by quantitative research based on questionnaires (possibly with post- or ante-interviews) on relevant societal segments, with well-defined statistical hypotheses, or in national organizations (or with multinational subsidiaries) to confirm and/or statistically complement the present qualitative correlations. The subsequent use of such qualitative, or quantitative, correlation helps in the development of competitive public (societal) scenarios and strategies, or business (organizational). The mutual advantages for the offer and demand in a complex competitive environment assure the sustainability (with an environmentally friendly impact) and is regenerative (renewed on the long-term spiral) for (meta-) economic generations too.

1.4 CONCLUSIONS AND PERSPECTIVES

The present study is a synthetic and original approach to the transcultural landscape, with its objective to provide a coherent and correlated image of societal, organizational, and individual cultures the latter, "melted" into the generational structure—for cognitive and methodological formalization. The study provides a fertile working base with transcultural dimensions, by promoting associated "eight-fold cultural dimensions," based on systematic combinations of three "core" values (the so-called trifecta point for each culture level); the study therefore introduces the 8D-intercultural matrices (Figs. 1.4 and 1.5) and apply them for generational–societal and generational–organizational phenomenological correlations for generation-X. The order of cultural dimensions in such a matrix is considered as "in the logic of scenarios," for example, in societal case, from communication to time (Fig. 1.4). However, at organizational level, the matrix dimensions goes from the rational truth to the strategy (of the organization), while capitalizing with the temporal dimension (because the strategy is in itself a long-term approach, so with temporal dynamics) (Fig. 1.5); in the same line of analysis, at generational level (the abscissas of Figs. 1.4 and 1.5), the dimensions cast from engagement with freedom in the middle, and culminating with motivation (precisely to "close the circle" and ensure continuity of existential cycles for the generation in question).

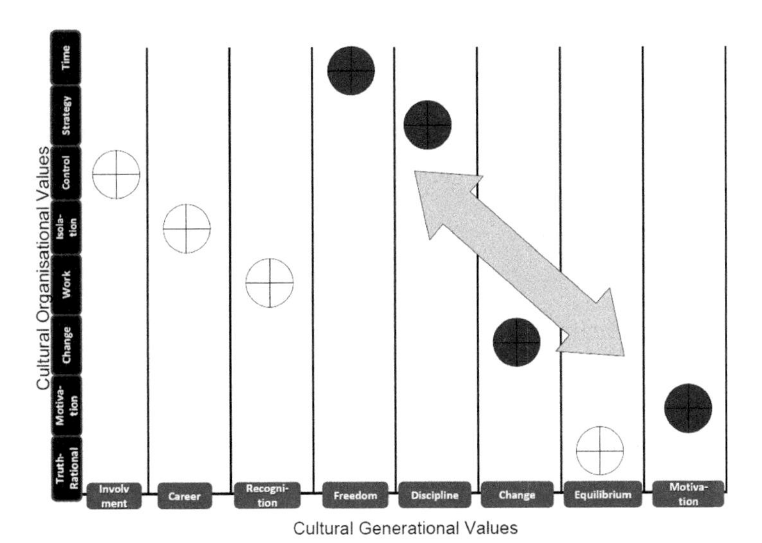

FIGURE 1.5 Intercultural matrix organization versus generation-X.

Of course, it can still be argued that these conclusions and correlations have a degree of arbitrariness, at least from two points of view:

- Lack of numerical support, quantitative, statistical;
- Lack of substantiating order in correlation matrices for the "dimensional inputs" specific from intercultural, societal–generational, and organizational–generational point of view.

Although for the first hypothetical objection above, the solution requires additional funds and logistic support for conducting field studies of sociological nature (given the group character of generations), the second hypothetical objection above should require a theoretical systematization based on cognitive correspondence of the coordinates of the cross-cultural trifecta points. Table 1.1 may give a basic line for such venture, that is, to imagine the points of a "cultural cub" being systematically generated, respectively (generational, organizational, societal); in this framework, the intercultural matrix analysis (generational–organizational, generational–societal) will be redone, considering the same "entry" order in the matrix for the cube points, as generated by the generational trifecta (a, b, and c); the same hierarchical position will display the same type of neighbors, so generating an algorithm, equally, to be applied to society and organizations as well (see the next chapters). Overall, such individual–organizational–societal constructs may be superimposed by various scaled cubes, so that the cognitive–behavioral–order curves can be detected for the intercultural crossing levels. The final picture will result in the desired (transcultural or interpersonal, or business organizational) networking (Putz, 2019).

KEYWORDS

- **eighth dimension**
- **trifecta (triple perfect) point**
- **individual**
- **organization**
- **society**
- **generation-X**
- **second culture**
- **comparative analysis and advantage**

REFERENCES

David, D. The Psychology of a Nation/People Between Kuhn and Torquemada: And Still It Is Moving (in Romanian Language). *Sinteza* (Romania) **2015,** *21*, 100–103.

Detert, J. R.; Schroeder, R. G.; Mauriel, J. J. A Framework for Linking Culture and Improvement Initiatives in Organizations. *Acad. Manage. Rev.* **2000,** *25* (4), 850–863.

Hofstede, G. *Culture's Consequences: International Differences in Work-Related Values*; Sage: London, 1980.

Hofstede, G. A European in Asia. *Asian J. Soc. Psychol.* **2007,** *10*, 16–21.

Hofstede, G. Dimensionalizing Cultures: The Hofstede Model in Context. *Online Read. Psychol. Culture* **2011,** *2* (1). http://dx.doi.org/10.9707/2307-0919.1014.

Hofstede, G.; Hosfstede, G. J.; Minkov, M. *Cultures and Organizations: Software of the Mind.* McGraw-Hill: New York, 2010.

Hofstede, G.; McCrae, R. R. Personality and Culture Revisited: Linking Traits and Dimensions of Culture. *Cross-Cultural Res.* **2004,** *38* (1), 52–88.

McCrae, R. R.; John, O. P. An Introduction to the Five-Factor Model and Its Applications. *J. Personality Social Psychol.* **1992,** *60*, 175–215.

Meyer, E. Navigating the Cultural Minefield. *Harvard Business Rev.* **2014**. https://hbr.org/2014/05/navigating-the-cultural-minefield.

Miller, G. A. The Magical Number Seven, Plus Or Minus Two: Some Limits On Our Capacity For Processing Information. *Psychol. Rev.* **1956,** *63*, 81–97.

Murphy, S. A. American Association of Retired People: Leading a Multigenerational Workforce, 2007. http://assets.aarp.org/www.aarp.org_/cs/misc/leading_a_multigenerational_workforce.pdf.

Nicolescu, B. *We, the Particle and the World* (in Romanian Language); Junimea: Iași, 2007.

Putz, M. V. *The Strategic Cube of the Distinctive Advantage: Epistemological Approach*, Chapter 2 (this monograph); 2019.

Robinson, W. S. Ecological Correlations and the Behavior of Individuals. *Am. Sociol. Rev.* **1950,** *15*, 351–357.

Williams, K. C.; Page, R. A. Marketing to the Generations. *J. Behav. Stud. Bus.* **2011,** *3*. http://www.aabri.com/jbsb.html.

CHAPTER 2

The Strategic Cube of the Distinctive Advantage: An Epistemological Approach

ABSTRACT

In the strategic management context, the strategic cube method is advanced toward unifying and integrating paradigm, for both classical and innovative methods, in reaching the competitive, sustainable, and regenerative advantage in qualitative and semiquantitative manner. In this regard, the algebraic and geometrical symmetry of matrix models specifies the cube (i.e., via transformations, as the rotations and reflections between the allowed strategic dots/states). The cube's strategic potential is illustrated by the comparative analysis of Blue Ocean and Red Ocean strategies, rediscovering the diagnostic methods of the strengths–weaknesses–opportunities–threats (SWOT), respectively, the strategic force model in analyzing the generic competition after Michael Porter. Results show the cubic method has internal consistency, while revealing the strategically specific predictions and distinction in the approached cases; at the same time, it opens the possibility of extending the strategic analysis on the 3D networks level in business organizations.

2.1 INTRODUCTION

In global modern economy, in civilized countries, the strategic management is the key component of general management, as the strategic management is developed at a superior level of management in order to achieve, implement, and maintain the competitive, sustainable, and regenerative advantage, so that the "recipe to success" of the company/organization/enterprise/corporation and even of nations (on the long-term vision) could not be easily imitated by competitors (on short-term mission) (Wright et al., 1989; Cockburn et al., 2000).

Furthermore, in order to develop a strategy, it is indispensable to have a complex algorithmic approach, a model as much as possible simple but not simplistic, comprehensively complex but not complicated, original or original reinvented of those already existing in the specialty literature (Petrişor, 2007). Algorithms, in their turn, can be conceptual–qualitative or/and numerical–quantitative; to have an appropriate evaluation–estimation–decision, they must have both components. And this brings us to numbers and to the fundamental question:

- Which is the ideal/optimal dimension, as number of variables, strategic attributes for a model/pattern (within strategic management) to be functional and significant?

One may speak about Porter's competitive pattern with its five or six forces (Porter, 2008), or on the chain of value (with five main activities), or recalling the strengths–weaknesses–opportunities–threats (SWOT) analysis (with the four quadrants), or on the political—economical–social–technological–ecological–juridical–cultural [PESTEcJ(C)] pattern with its seven components, or even on the Ansoff matrix (with its two developmental axes and four specializations, so resulting into 2 × 4 phenomenological entrances), or arriving at the celebrated Boston Consulting Group (BCG) matrix 2 × 2 entrances, or on its ABC matrix variant of McKensey with 3 × 3 = 9 phenomenological entrances, and so on. However, which is the optimal work number with dimensions and strategic variables in the strategic management? And also,

- Is there a number and, respectively, a unifying pattern for "classical" patterns of strategic management able to integrate all of them and possibly reinvent them at a methodological–cognitive dimension to finally have a superior strategic impact?

The present study aims to deal with such a bold approach!

2.2 COGNITIVE ANALYSIS

We start our attempt to research the "magical number of strategic management by going back to the founders of management, Fayol versus Taylor, and we end up by noticing something quite interesting as follows:

- Frederick Winslow Taylor's first work (1903), Shop Management, recommends that in order to achieve economic effectiveness, a division of tasks for the first-line boss into eight specializations is needed (see Drucker's classification below). This finally led to the implementation and consecration of management through objectives, specific to economic activities in American companies;
- Taylor's study was translated into French in 1913 and was carefully analyzed by Henri Fayol who allocated a special space in his classical book "Administration industrielle et generale" [*Industrial and General Administration*] (Fayol 1916), where he criticized the too many dimensions allocated, suggesting instead one authority in the person of the manager. The implementation rules are optimal ones stressing on efficiency.

One remarks the distinction between the one-dimensional operational management (Fayol) and the strategic multidimensional one (Taylor).

The modern father of management, Peter Drucker (1954, republished 2006) establishes eight key domains to perform management through objectives (and not only) which should define all the compartments of an organization/economic entity, as it follows (Bibu et al., 2008):

- Market standing: setting objectives/goals in relation to results/objectives of the competitors.
- Innovation: setting objectives/goals so that they should include improvement of current products and services, with at least 20% within a 3-year cycle at least.
- Productivity: setting objectives/goals which include the enhancement of productivity standards for all operational areas (e.g., 8% growth of production with the same staff, 12% of growth of efficiency of installations in use, etc.).
- Physical and financial resources: setting specific objectives related to the use, acquisition, and maintenance of capital and monetary

resources (e.g., investing capital to increase the capacity to provide services for the clients, reduce long-term debts, identify cheaper suppliers and materials able to reduce acquisition and production costs, etc.).

- Profitability: setting minimum levels of performance of the company (e.g., increase the profit rate in the years to come, etc.).
- Managerial performance and development: setting objectives in order to develop operational management talent (on short term) and strategic management (on long term), for example, by setting a performance evaluation system for all employees in a certain year, attendance to training courses, etc.
- Worker performance and attitude: setting new work standards for the employees with no management positions (e.g., reduce the absences without leave, increase the working hours annually, etc.).
- Public and social responsibility: it is directly related to the durable and sustainable character of the company/organization's activity; it defines the minimum degree of satisfaction of the needs and interests of the company it serves and/or belongs to; with eminent examples such as: protection of the environment activities, sponsorship programs, programs for the use of the community, etc.

Further on, the sociologist Alvesson (2002) identifies eight metaphors of the organizational culture as follows (Hofstede et al. (2010), translated to Romanian 2012):

- Mechanism control for an informal contract
- Compass which indicated direction of priorities
- Social binder to identify with the company
- Holy cow that people worship
- Affective regulator of emotions and their expression
- Mix of conflicts, ambiguities, and fractions
- Ready-made ideas which lead to a dead end
- Closed system of ideas and meanings, which prevent people from exploring new opportunities.

Even more, while being oriented either to life, society, or environment, as the strategic postmanagement should be, Lorenz (1973)—Nobel prize for medicine and physiology—already distinguished eight capital "sins," we would rather call them dangers (lethal, or at least harmful on long term, thus having a strategic impact) for the contemporary and further society, specific to civilized world already situated in a postmodern era, namely:

- Overpopulation—not to get emotionally involved
- Devastation of the vital space (unbalanced hunt and pray); illegal exploitation, loss of information, and ecology
- Competing with oneself: intraspecific selection, reflection, fear, technological development, and utilitarianism
- Heat death of senses: market state effort—pleasure, joy, comfort, the sin of drowsiness, development of pharmacology, natural obstacle, and neophilia (reinforcement, instant gratification, and built-in obsolation)
- Genetic decay: oscillations, inertia, true values, natural sense of right, hospitalization
- Crashing tradition: cumulative tradition, homo faber, uniform, national hatred, foreign ethnical group, and creative reconstruction
- Receptiveness to indoctrination: falsifying hypotheses, per exclusionem, reflex and conditioning, uniformity of mentalities, tabula rasa, *Homo sapiens*, infinite structure, objective research, psychology void of soul, dehumanization affecting science, and steps abroad (a rather simple extension of physics, Crick op. cit.)
- Nuclear weapons: infinite atmosphere of the world.

What does this number 8 mean? From the point of view of numerology, it is the only number that symbolically covers all the other numbers (including zero and infinite) only through decomposing it. Alternatively, one may say that 8 is "giving up" its valences (from 1 to 9, 0, and infinity), and not by adding something (from "nothingness"); it is worth observing that the last case is the case of any other numbers in the attempt to reconstruct all the others (Table 2.1).

TABLE 2.1 "Creative" Valences of Number 8 by Adapted Renunciation of the 7 "Intrinsic Valences" or by Reorganizing Them.

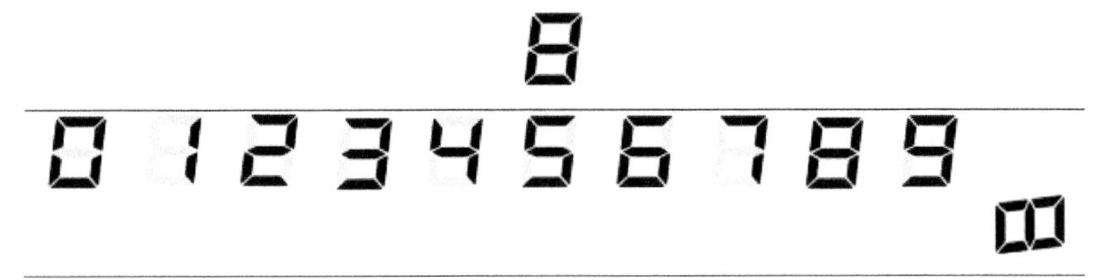

Number 8 corresponds to a musical octave, not accidentally associated by Pythagoras to the "music of spheres," (Fig. 2.1) to the Earth—"center of the Universe" (James, 1995):

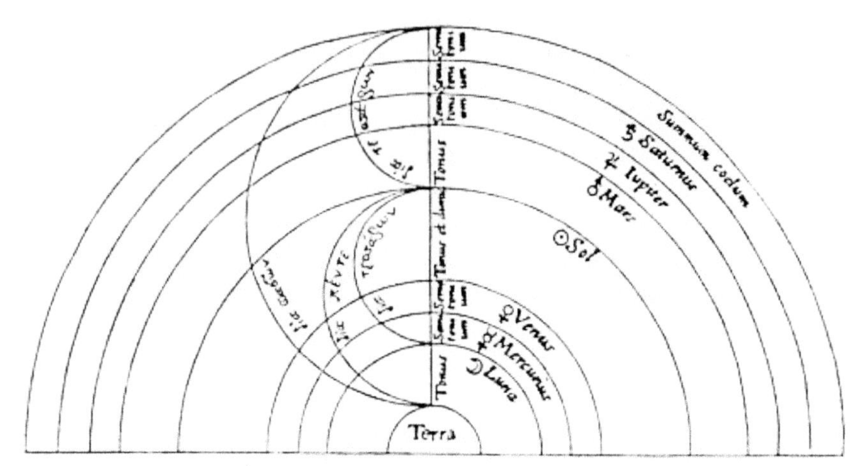

FIGURE 2.1 Pythagoras music of spheres.

Source: Adapted with permission from Stanley (1656).

Number 8 is also associated to the natural spectrum of colors: 7 ROGVAIV "valences" + "origin" at the end (IR-infrared) or in the opposite extremity ultraviolet.

We partially conclude that it can be rationalized to the golden triplet of the economic approach: generating, promoting, and maintaining the distinctive advantage of an organization/business/product/service as being:

- competitive (diverse and/or different),
- sustainable (useful from a social point of view and environmental friendly, multigenerationally concern),
- and regenerative (reborn, from time to time, eventually in strategic cycles)

by strategic development on eight dimensions/orientations, yet to be identified, by giving sense (mathematic and numerical) and significance (quantitative and decisional) in a coherent manner both in substantiation and as an analysis instrument.

Thus, the cognitive potential of the cube in the strategic management of the distinctive advantage is defined in the present approach (algebraic) as follows:

It is considered the triplet of the distinctive advantage in the algebraic form of line vectors:

- competitive advantage $C = (1 \quad 0 \quad 0)$;
- sustainable advantage $S = (0 \quad 1 \quad 0)$;
- regenerative advantage $R = (0 \quad 0 \quad 1)$;

By combining the three line vectors, we obtain the matrix 3 × 3 of the competitive, sustainable, and regenerative advantage, practically representing the trifecta point (three times perfect) in the development of a business/organization/product/economic process, etc.

$$CSR = \begin{pmatrix} 1 & 0 & 0 \\ 0 & 1 & 0 \\ 0 & 0 & 1 \end{pmatrix} \quad (2.1)$$

Analogously, we can build through the symmetry of negation:

- competitive disadvantage $\overline{C} = (-1 \quad 0 \quad 0)$;
- sustainable disadvantage $\overline{S} = (0 \quad -1 \quad 0)$
- regenerative disadvantage $\overline{R} = (0 \quad 0 \quad -1)$;

resulting in the matrix 3 × 3 of the competitive, sustainable, and regenerative disadvantage or the matrix of the noncompetitive, nonsustainable, and nonregenerative advantage:

$$\bar{C}\bar{S}\bar{R} = \begin{pmatrix} -1 & 0 & 0 \\ 0 & -1 & 0 \\ 0 & 0 & -1 \end{pmatrix} \tag{2.2}$$

Besides the forms CSR and $\bar{C}\bar{S}\bar{R}$, we can also obtain intermediate combinations, such as:

- derived forms of the matrix CSR by successive negation of one trifecta component

$$\bar{C}SR = \begin{pmatrix} -1 & 0 & 0 \\ 0 & 1 & 0 \\ 0 & 0 & 1 \end{pmatrix}, \tag{2.3}$$

$$C\bar{S}R = \begin{pmatrix} 1 & 0 & 0 \\ 0 & -1 & 0 \\ 0 & 0 & 1 \end{pmatrix}, \tag{2.4}$$

$$CS\bar{R} = \begin{pmatrix} 1 & 0 & 0 \\ 0 & 1 & 0 \\ 0 & 0 & -1 \end{pmatrix}; \tag{2.5}$$

- derived forms of the matrix CDP by successive negation of two trifecta components

$$\bar{C}\bar{S}R = \begin{pmatrix} -1 & 0 & 0 \\ 0 & -1 & 0 \\ 0 & 0 & 1 \end{pmatrix}, \tag{2.6}$$

$$\bar{C}S\bar{R} = \begin{pmatrix} -1 & 0 & 0 \\ 0 & 1 & 0 \\ 0 & 0 & -1 \end{pmatrix}, \tag{2.7}$$

$$C\bar{S}\bar{R} = \begin{pmatrix} 1 & 0 & 0 \\ 0 & -1 & 0 \\ 0 & 0 & -1 \end{pmatrix}; \tag{2.8}$$

- there are no other derived forms.

It is quite interesting how points of the distinctive advantage form an algebraic group in relation with the matrix multiplication ⊗. In other words, any two "strategic points," or the above strategic matrixes, when multiplied together, result into a matrix existing between points (1)–(8), see Annex 2.1, without generating any additional matrix to the ones already identified.

The methodological advantage of the present algebraic approach resides in the capacity of the above (1)–(8) points "practically close" the strategic space of successive combinations between any points of the cube (thus having become strategic itself) (Fig. 2.2). Moreover, they can be dynamically combined by coupling one point to the other through the symmetry transformations allowed to the cube (see Annex 2.2 and the application of the next section). It is worth mentioning that for the representation in Figure 2.2, the diagonal matrixes (1)–(8) are interpreted in the 3D sense by "reading on the main diagonals of the Cartesian coordinates (0x, 0y, 0z) of the identified strategic points, that is:

- $CSR = (1,1,1)$; $\bar{C}\bar{S}\bar{R} = (-1,-1,-1)$,
- $\bar{C}SR = (-1,1,1)$; $C\bar{S}R = (1,-1,1)$; $CS\bar{R} = (1,1,-1)$;
- $\bar{C}\bar{S}R = (-1,-1,1)$; $\bar{C}S\bar{R} = (-1,1,-1)$; $C\bar{S}\bar{R} = (1,-1,-1)$

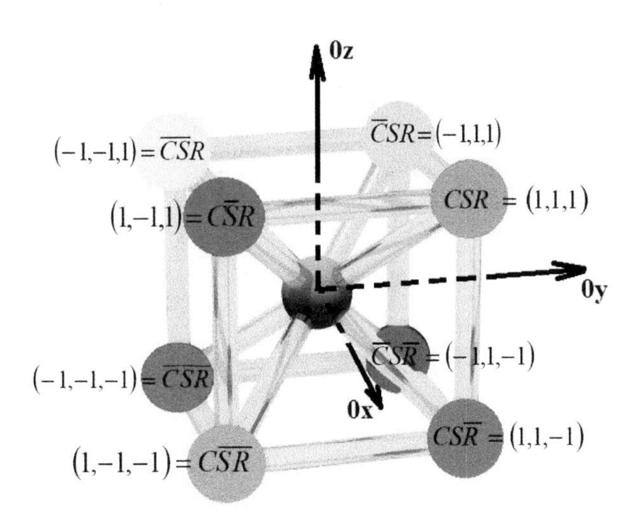

FIGURE 2.2 (See color insert.) The strategic cube with strategic points (1)–(8) interpreted in Cartesian coordinates 3D in relation to the central point of coordinates (0,0,0).

It is only to establish the significance of the strategic points in the strategic cube of Figure 2.2.

To do this, the senses of the 8× of the introductive levels are to be combined (see Introduction of this chapter):

- management through objectives (Drucker, 1954),
- organizational culture (Alvesson, 2002),
- capital sins of the civilized world (Lorenz, 1973).

Settled in a more or less cognitive (society and natural environment), linguistic (community–organization), anthropocentric (individual and his generation) deconstructive process, in the sense of Jacques Derrida's (2000) creative deconstruction, thus being perfectible (Petrişor, 2007; Akerlof and Shiller, 2010), one may suggest the following ("colored") attributes:

- For the strategic point CSR, we obtain the strategic attribute: wise business
- For the strategic point $\bar{C}\bar{S}\bar{R}$, we obtain the strategic attribute: animal spirits
- For the strategic point $\bar{C}SR$, we obtain the strategic attribute: klepto-cratic underground business
- For the strategic point $C\bar{S}R$, we obtain the strategic attribute: profit-able business
- For the strategic point $CS\bar{R}$, we obtain the strategic attribute: smart business
- For the strategic point $\bar{C}\bar{S}R$, we obtain the strategic attribute: perverse spirits
- For the strategic point $\bar{C}S\bar{R}$, we obtain the strategic attribute: outsourcing
- For the strategic point $C\bar{S}\bar{R}$, we obtain the strategic attribute: polemocracy

Together, these strategic points/settings generate the actual strategic cube in Figure 2.3; each point has the cognitive potential established by its position and attribute in relation to the others. Accordingly, each point taken as a reference can combine all others, and even can be transformed into them through allowed symmetries in compliance with the cubic symmetry elements (symmetry axes and planes) to which the reference point can be related (Annex 2.2). It results what could be called the strategic map of transformation potentials of the distinctive advantage in the strategic cube specific to the approached strategic pattern—as it will be illustrated in the next section.

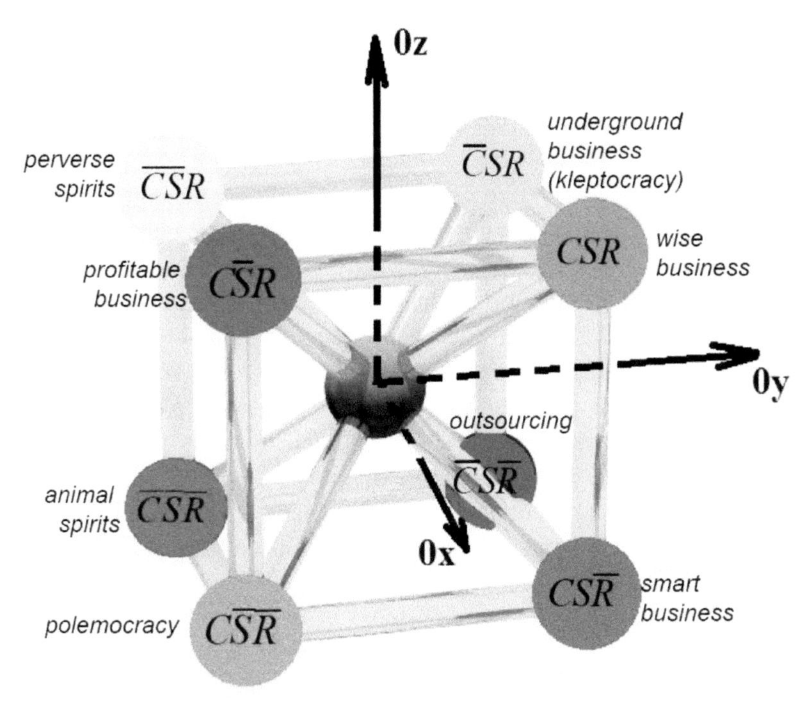

FIGURE 2.3 (See color insert.) The strategic cube with the strategic attributes corresponding to strategic points in Figure 2.2.

2.3 METHODOLOGICAL ANALYSIS

We can now use the mathematical (and cognitive) pattern to settle, qualitatively and semiquantitatively for now (Putz, 2019), the strategic potential of the cube in the strategic management, by following the "rule of the maximum of symmetries in the cube":

- It is embedded, integrated, and circumscribed a strategic pattern given or newly created in the strategic cube in Figure 2.2, so that the maximum of symmetries are kept in rotations and reflection specific to the cube.

- The rotations that remained valid define the strategic dynamics and, respectively, the strategic potential of the analyzed pattern in relation to the cube symmetries.
- The various strategic patterns can be integrated–embedded in the strategic cube according to the rule of maximal symmetries, resulting that the pattern with a higher strategic potential is the one that allows more "strategic movements" with the symmetries of the strategic cube it belongs to.

Consequently, the pattern of the cube of strategic management implies the processes associated with the symmetries of the cube (Annex 2.2) (Putz, 2017).

Furthermore, the method of the potential of the strategic cube will be applied in order to perform the phenomenological analysis (qualitative) of the strategic management specific to the Blue Ocean versus Red Ocean, in the original view of its authors (Kim and Mauborgne, 2014, 2015).

Blue Ocean strategy implies, creates, and involves new product and services markets. In this respect, the diagnosis map with related strategic action/development potential is originally adapted here onto the paradigm of the SWOT analysis in an adapted form of points/sectors/actions to be increased–eliminated–created–reduced, respectively, as it is illustrated in Table 2.2 (Kim and Mauborgne, 2014).

Transposing the matrix of Blue Ocean of Table 2.2 into the strategic cube of Figure 2.3 generates the cube Blue Ocean of Figure 2.4, for which we identify the following symmetry dynamics, based on combining the symmetry elements identified in the strategic cube of Annex 2.2 and the strategic cube group of Annex 2.1:

- First symmetry axis A^3 (axis $CS\bar{R} \leftrightarrow \bar{C}\bar{S}R$)
    - first rotation: $\bar{C}SR \times C\bar{S}\bar{R} = \bar{C}\bar{S}\bar{R}$ ~animal spirits (Red Ocean)
    - second rotation $C\bar{S}R \times \bar{C}S\bar{R} = \bar{C}\bar{S}\bar{R}$ ~animal spirits (Red Ocean)

- Second symmetry axis A^3 (axis $\bar{C}SR \leftrightarrow C\bar{S}\bar{R}$)
    - the only first rotation $C\bar{S}R \times \bar{C}S\bar{R} = \bar{C}\bar{S}\bar{R}$ ~ animal spirits (Red Ocean)

- Third symmetry axis A^3 (axis $C\bar{S}R \leftrightarrow \bar{C}S\bar{R}$)
    - the only rotation $\bar{C}SR \times C\bar{S}\bar{R} = \bar{C}\bar{S}\bar{R}$ ~ animal spirits (Red Ocean)

TABLE 2.2 Analysis Matrix for Blue Ocean Strategy, Particularized at the Level of Medium Management, with the Motto "More Coaching, Less Control."

Strengths (to increase)	**Opportunities (to create)**
Actions and activities regarding evaluation of the current management reality which are to be improved–amplified (in time and in a smart way), for example, to increase the positive learning environment, clear explanation of the organization's strategy, freedom of first-line managers/operational to increase decisional flexibility, coaching of the personnel.	Actions and activities regarding institutionalization of new management practices to be considered (in time and in a smart way); they are not currently implemented in the organization–yet in need to be created as a set of common performance objectives, sharing good practices with all work teams, and aligning benefits to performance.
Weaknesses (to eliminate)	**Threats (to reduce)**
Actions and activities regarding development of an alternative management profile which has to be eliminated (in time and in a smart way), for example, (to eliminate) more than one person responsible for the same initiative, frequent request of detailed reports on the progress of an initiative	Actions and activities regarding selection of the optimal management profile to be considered (in time and in a smart way) to be diminished and finally eliminated as uncompetitive (to reduce) time spent with top managers, request and analysis of justifications for inferior management decisions, mandatory fulfillment of specific requests for business activities

Source: Adapted with permission from Kim and Mauborgne (2014).

- Fourth symmetry axis A^3 (axis $CSR \leftrightarrow \overline{CSR}$)
  - first rotation $\bar{C}SR \times C\overline{\bar{S}R} = \overline{\bar{C}SR}$ ~ animal spirits (Red Ocean)
  - second rotation $C\bar{S}R \times \bar{C}S\bar{R} = \overline{\bar{C}S\bar{R}}$ ~ animal spirits (Red Ocean)
- First symmetry axis A^2 (on 0y negative):
  - first rotation : $\bar{C}SR \times \bar{C}S\bar{R} = CS\bar{R}$ ~smart business;
  - second rotation: $C\bar{S}R \times C\overline{\bar{S}R} = CS\bar{R}$ ~ smart business;
- Second symmetry axis A^2 (on 0y positive):
  - first rotation : $\bar{C}SR \times C\overline{\bar{S}R} = \overline{\bar{C}SR}$ ~ animal spirits (Red Ocean)
  - second rotation: $C\bar{S}R \times \bar{C}S\bar{R} = \overline{\bar{C}S\bar{R}}$ ~ animal spirits (Red Ocean)
- The only symmetry plan that requests change m ($\perp 0z$):
  - first reflection : $C\bar{S}R \times C\overline{\bar{S}R} = CS\bar{R}$ ~ smart business;
  - second reflection $\bar{C}SR \times \bar{C}S\bar{R} = CS\bar{R}$ ~ smart business.

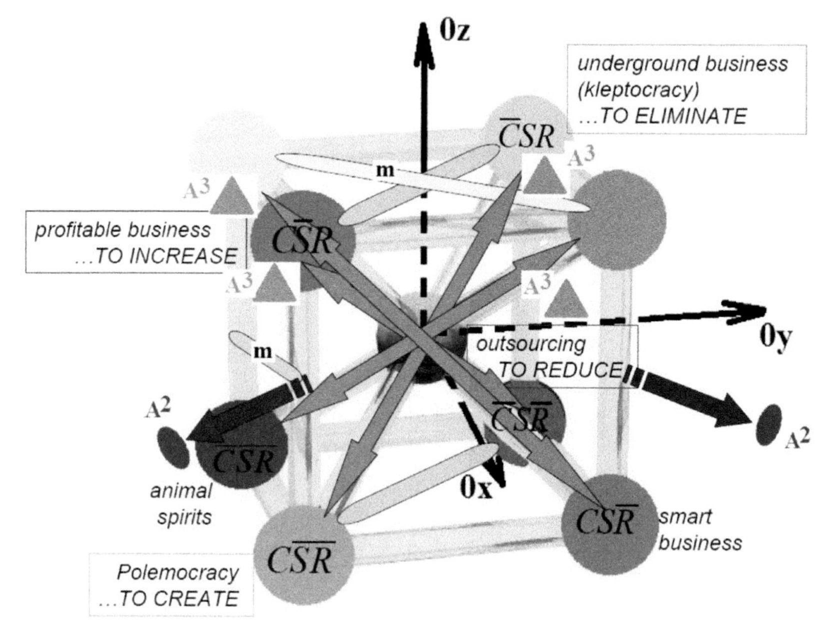

FIGURE 2.4 Blue Ocean strategic cube abstracted from Figure 2.3, according to matrix of Table 2.2.

The partial conclusion comes with a notice: the Blue Ocean strategy can generate in the dynamics both "the decadence and the strategic degradation," alike; however, a slide to the Red Ocean can also generate a smart business through alternating the nonsustainable but regenerative competitive advantage (profitable business) with the sustainable and nonregenerative competitive advantage (smart business). A similar study can be redone when it is considered that the four input points/states/strategic positions being associated to combinations:

$$\{C\bar{S}R, CSR, CS\bar{R}, C\bar{S}\bar{R}\},\ \{C\bar{S}R, CSR, \bar{C}SR, \bar{C}\bar{S}R\},\ \{C\bar{S}R, \bar{C}\bar{S}R, C\bar{S}\bar{R}, \bar{C}\bar{S}\bar{R}\}.$$

Red Ocean strategy supposes on the other hand avoiding the so-called "strategic traps" in reaching the distinctive advantage (competitive, sustainable, and regenerative), according to Kim and Mauborgne, 2015:

- Trap 1: The strategic approach of creating market based on customer/consumer-oriented approach (those existing), for example, Sony (with the program portable reader system—PRS released in 2006) versus Amazon (with the program Kindle released in 2006) regarding the e-books system, a war lost by Sony.
- Trap 2: The niche strategic approach in creating a market, for example, the mistake of the Delta company by releasing in 2003 the low-cost program, as too much targeted on the market segment destined to career women—so the program was closed in April 2006, 36 months after its release; on the other hand, Pret-A-Manger was a successful initiative by creating the strategy (by market desegmentation) for quality sandwich restaurant at affordable prices.
- Trap 3: Mistaking technological innovation for creation of a market strategy, for example, the wine company Yellow Tail, the coffee company Starbucks, or the entertainment company Cirque du Soleil successfully found new markets without making appeal to modern technology; in exchange, the Segway personal transporter (on two wheels) failed despite the high tech implementations in the products because of the relatively high cost (around 5000 USD) for promoting means of transport with urban inconveniences (with difficult portability on the underground train, low parking security, difficult off-road driving, etc.).
- Trap 4: Mistaking destructive/disruptive creation for market strategy creation, for example, the successful story of Viagra which introduced on the market a new lifestyle without destroying the existing one related to intimate life; the same with the creation of Grameen Bank which introduced microfinancing industry without

affecting the general banking system; likewise, with the Nintendo Wii product for video games was appealing for both adult segment and the one dedicated to children and teenagers.

- Trap 5: Mistaking differentiation for market strategy creation, for example, forcing "the productivity frontiers" by strategic differentiation (offering premium products) is not always an option: such mistake was made by BMW by releasing the scooter motorbike C1 in 2000 destined to ensure safety and efficiency in the traffic—but with a very high production cost and an acquisition price of 7000–10,000 USD, twice bigger than the market price; thus, BMW had to stop the production in 2003.
- Trap 6: Mistaking the strategy of low price for market strategy creation; it is the reverse of the mistake in Trap 5: a mistake made by Ouya company in launching video games in 2013, several times cheaper (approx. 99 USD) than the similar games of the competitors on the market, such as Sony, Microsoft, Nintendo (with prices between 199 and 419 USD)—so they still failed because they lacked a portfolio of high quality games (featuring 3D intensity, enhanced graphics, cell phone command—the so-called play-on-the-go features), soon arriving in selling process to the shareholders.

These "traps" of the Red Ocean can be associated (in an original approach here) to the extended pattern of Porter's 5 + 1 forces, as it is illustrated in Figure 2.5.

Transposition of Porter's Forces of Red Ocean of Figure 2.5 into the strategic cube of Figure 2.3 generates the cubic Red Ocean of Figure 2.6. It identifies a dynamics specific to symmetry, as based on combining symmetry elements identified in the strategic cube (Annex 2.2) and on the group of the strategic cube (Annex 2.1), so it lists:

- Symmetry axes A^3 are identical, in the selected configuration, to those of the analysis of Blue Ocean and invariably results into maintaining the Red Ocean strategy/low price (F1 Porter), without bringing added conceptual–strategic value;
- First symmetry plan that requests change, m (includes competitiveness axis):
    - first reflection: $F2 \times F1 = \bar{C}SR \times \bar{C}\overline{S}\overline{R} = C\overline{S}\overline{R}$ ~ F6/polemocracy;
    - second reflection $F3 \times F6 = CSR \times C\overline{S}\overline{R} = C\overline{S}\overline{R}$ ~ F6/polemocracy;

- Second symmetry plan that requests change, *m* (includes durability axis):
    - first reflection: $F2 \times F5 = \bar{C}SR \times CS\bar{R} = \bar{C}S\bar{R}$ ~ outsourcing;
    - second reflection $F4 \times F6 = \bar{C}\bar{S}R \times C\bar{S}\bar{R} = \bar{C}S\bar{R}$ ~ outsourcing;
- Third symmetry plan that requests change *m* (includes regenerative axis):
    - first reflection: $F1 \times F5 = \overline{CSR} \times CS\bar{R} = \bar{C}\bar{S}R$ ~ F4/ perverse spirits;
    - second reflection $F3 \times F4 = CSR \times \bar{C}\bar{S}R = \bar{C}\bar{S}R$ ~ F4/ perverse spirits.

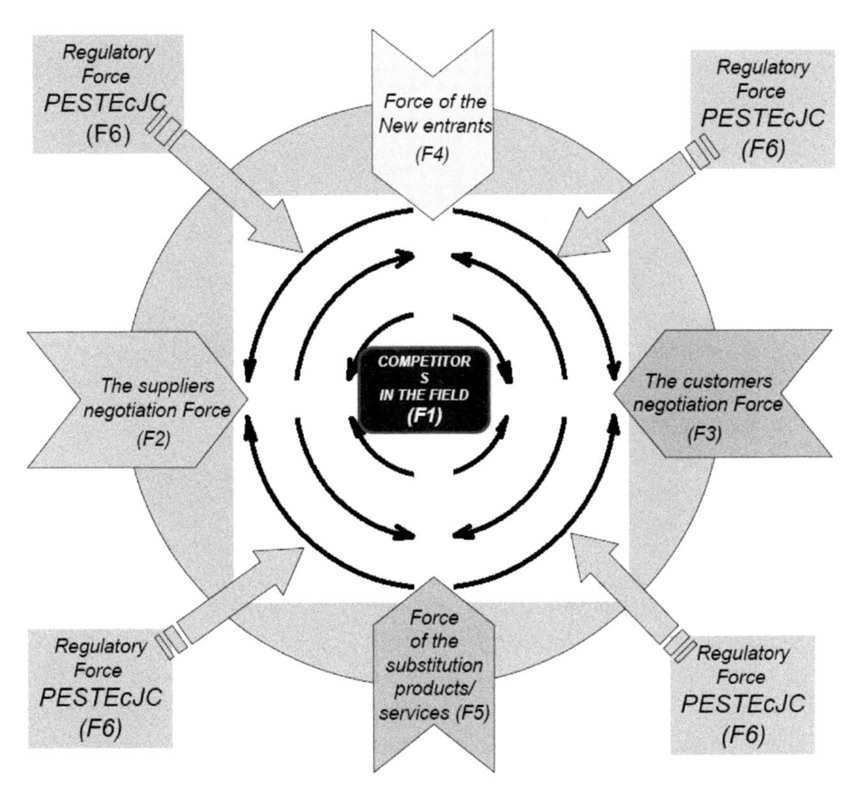

FIGURE 2.5 Porter's 5 + 1 forces specific to Red Ocean; to be mentioned that the sixth force refers to the PESTEcJC influence specific to the environment it works in refers to the competitive domain in question.

It is noticed that in the Red Ocean strategy, through the dynamics of the strategic cube, it is also obtained a strategy which is external to the Porter Forces, by identifying the outsourcing, possibly considered the seventh force in an extended diagram. What is interesting is that from no dynamic combination in the Red Ocean strategy can we get the distinctive status of Blue Ocean, that is, the status of profitable business!

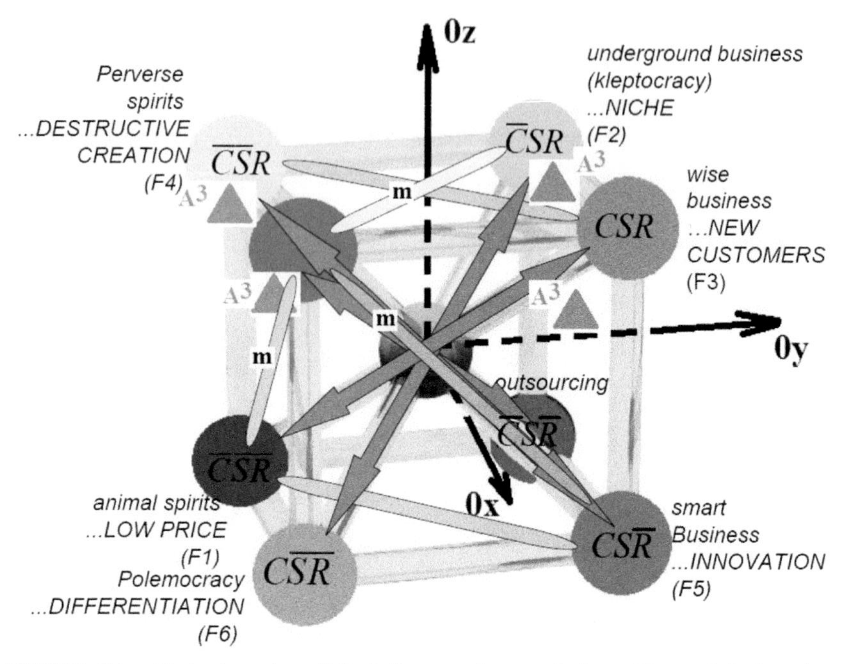

FIGURE 2.6 Strategic cube of Red Ocean, abstracted from Figure 2.3, based on identification of the specific Porter forces 5 + 1 of Figure 2.5.

Thus, the analysis of the strategic cube can reveal how the Blue Ocean can slide or can degrade into Red Ocean. Yet, it is not possible to create Blue Ocean from Red Ocean without a change of paradigm, respectively, without input strategic states, which corresponds to the total reconfiguration of the organization, in Chandlerean sense, from structure to performance and then to competitiveness, in this way avoiding the "Red" corner $\overline{CSR}$.

2.4 CONCLUSIONS AND PERSPECTIVES

The present approach of strategic management finds itself in a stage of reformation, in parallel with current and future global economic changes and social challenges. In this context, it is imperiously necessary to reform the paradigm of strategic analysis. It aims at finding a unitary approach, an ideal–universal one, in order to correlate and integrate the existing management methods, techniques, and practices while passing from an operational level to a strategic one, from organizations to interested parties toward a global inclusive economic and social network. The present study makes a step forward in fundamental research of strategic management, by suggesting the use of algebra and geometry in approaching the strategic attributes such as the distinctive advantage, competitiveness, sustainability, regenerative. They are further projected into a symmetric space, highly informational, inter-relational while still open to later couplings. The cubic shape ingeniously offers such framework by its inter-related eight states—all generated from the basic trifecta point of the distinctive advantage. Epistemological considerations, cognitive and methodological ones are brought, in an integrated manner and are applied to analysis/strategic diagnose methods SWOT and that of Porter's Forces, accompanying the strategic analysis for Blue and respectively Red Ocean. The results are relevant, as far as the inner coherence of the method, the fundamentally strategic predictions, the directed confirmation (injective and not bijective) between the Blue and Red Oceans in the strategic cube are concerned. The method also allows later extensions and elaborations to the approaches of neural economy, in the postmodern strategic management (Putz, 2017, 2019).

ANNEX 2.1 THE GROUP OF THE STRATEGIC CUBE

For the matrixes of the distinctive advantage (1)–(8) the direct multiplying operations are formed as follows:

- For matrix $\boxed{CSR}$, eq 2.1, in turn multiplied by all the others, we reobtain all the other points as we deal with multiplying by the unity matrix:

$$CSR \otimes CSR = \begin{pmatrix} 1 & 0 & 0 \\ 0 & 1 & 0 \\ 0 & 0 & 1 \end{pmatrix} \begin{pmatrix} 1 & 0 & 0 \\ 0 & 1 & 0 \\ 0 & 0 & 1 \end{pmatrix} = \begin{pmatrix} 1 & 0 & 0 \\ 0 & 1 & 0 \\ 0 & 0 & 1 \end{pmatrix} = CSR; \qquad \text{(A1.1-1)}$$

$$CSR \otimes \overline{CSR} = \begin{pmatrix} 1 & 0 & 0 \\ 0 & 1 & 0 \\ 0 & 0 & 1 \end{pmatrix} \begin{pmatrix} -1 & 0 & 0 \\ 0 & -1 & 0 \\ 0 & 0 & -1 \end{pmatrix} = \begin{pmatrix} -1 & 0 & 0 \\ 0 & -1 & 0 \\ 0 & 0 & -1 \end{pmatrix} = \overline{CSR}; \quad \text{(A1.1-2)}$$

and similarly for other strategic points:

$$CSR \times \bar{C}SR = \bar{C}SR; \quad \text{(A1.1-3)}$$

$$CSR \times C\bar{S}R = C\bar{S}R; \quad \text{(A1.1-4)}$$

$$CSR \times CS\bar{R} = CS\bar{R}; \quad \text{(A1.1-5)}$$

$$CSR \times \overline{CS}R = \overline{CS}R; \quad \text{(A1.1-6)}$$

$$CSR \times \bar{C}S\bar{R} = \bar{C}S\bar{R}; \quad \text{(A1.1-7)}$$

$$CSR \times C\overline{SR} = C\overline{SR}. \quad \text{(A1.1-8)}$$

For matrix $\boxed{\overline{CSR}}$ in turn multiplied by all the others, we get respectively:

$$\overline{CSR} \times CSR = \begin{pmatrix} -1 & 0 & 0 \\ 0 & -1 & 0 \\ 0 & 0 & -1 \end{pmatrix} \begin{pmatrix} 1 & 0 & 0 \\ 0 & 1 & 0 \\ 0 & 0 & 1 \end{pmatrix} = \begin{pmatrix} -1 & 0 & 0 \\ 0 & -1 & 0 \\ 0 & 0 & -1 \end{pmatrix} = \overline{CSR}; \quad \text{(A1.2-1)}$$

and similarly for other strategic points. One notices the “multiplication with negation” rule which results into the “positive value” for multiplying two negations and, respectively, the “negative value” result by multiplying positive and negative or negative with positive points’ signs, without calculating the matrix product. Moreover, their diagonal shapes ensure the “noncrossing,” or more precisely the “independence” of 3D attributes (respectively, along the lines and columns specific to competitiveness, sustainable and regenerative); all these contributed to the methodological advantage of the present algebraic approach. Thus we obtain, in a direct manner

$$\overline{CSR} \times \overline{CSR} = \begin{pmatrix} -1 & 0 & 0 \\ 0 & -1 & 0 \\ 0 & 0 & -1 \end{pmatrix} \begin{pmatrix} -1 & 0 & 0 \\ 0 & -1 & 0 \\ 0 & 0 & -1 \end{pmatrix} = \begin{pmatrix} 1 & 0 & 0 \\ 0 & 1 & 0 \\ 0 & 0 & 1 \end{pmatrix} = CSR; \quad \text{(A1.2-2)}$$

$$\overline{CSR} \times \bar{C}SR = C\overline{SR}; \quad \text{(A1.2-3)}$$

$$\overline{CSR} \times C\bar{S}R = \bar{C}S\bar{R}; \quad \text{(A1.2-4)}$$

$$\bar{C}\bar{S}\bar{R}\times CS\bar{R}=\bar{C}\bar{S}R; \qquad \text{(A1.2-5)}$$

$$\bar{C}\bar{S}\bar{R}\times \bar{C}\bar{S}R=CS\bar{R}; \qquad \text{(A1.2-6)}$$

$$\bar{C}\bar{S}\bar{R}\times \bar{C}S\bar{R}=C\bar{S}R; \qquad \text{(A1.2-7)}$$

$$\bar{C}\bar{S}\bar{R}\times C\bar{S}\bar{R}=\bar{C}SR. \qquad \text{(A1.2-8)}$$

The same rule for the other strategic products:

- For matrix $\boxed{\bar{C}SR}$ multiplied in turns with the others, we get, respectively:

$$\bar{C}SR\times CSR=\bar{C}SR; \qquad \text{(A1.3-1)}$$

$$\bar{C}SR\times \bar{C}\bar{S}\bar{R}=C\bar{S}\bar{R}; \qquad \text{(A1.3-2)}$$

$$\bar{C}SR\times \bar{C}SR=CSR; \qquad \text{(A1.3-3)}$$

$$\bar{C}SR\times C\bar{S}R=\bar{C}\bar{S}R; \qquad \text{(A1.3-4)}$$

$$\bar{C}SR\times CS\bar{R}=\bar{C}S\bar{R}; \qquad \text{(A1.3-5)}$$

$$\bar{C}SR\times \bar{C}\bar{S}R=C\bar{S}R; \qquad \text{(A1.3-6)}$$

$$\bar{C}SR\times \bar{C}S\bar{R}=CS\bar{R}; \qquad \text{(A1.3-7)}$$

$$\bar{C}SR\times C\bar{S}\bar{R}=\bar{C}\bar{S}\bar{R}. \qquad \text{(A1.3-8)}$$

- For matrix $\boxed{C\bar{S}R}$ multiplied in turns with the others, we get, respectively:

$$C\bar{S}R\times CSR=C\bar{S}R; \qquad \text{(A1.4-1)}$$

$$C\bar{S}R\times \bar{C}\bar{S}\bar{R}=\bar{C}S\bar{R}; \qquad \text{(A1.4-2)}$$

$$C\bar{S}R\times \bar{C}SR=\bar{C}\bar{S}R; \qquad \text{(A1.4-3)}$$

$$C\bar{S}R\times C\bar{S}R=CSR; \qquad \text{(A1.4-4)}$$

$$C\bar{S}R\times CS\bar{R}=C\bar{S}\bar{R}; \qquad \text{(A1.4-5)}$$

$$C\bar{S}R\times \bar{C}\bar{S}R=\bar{C}SR; \qquad \text{(A1.4-6)}$$

$$C\bar{S}R\times \bar{C}S\bar{R}=\bar{C}\bar{S}\bar{R}; \qquad \text{(A1.4-7)}$$

$$C\bar{S}R\times C\bar{S}\bar{R}=CS\bar{R}. \qquad \text{(A1.4-8)}$$

- For matrix $\boxed{CS\bar{R}}$ multiplied in turns with the others, we get, respectively:

$$CS\bar{R}\times CSR = CS\bar{R}; \tag{A1.5-1}$$

$$CS\bar{R}\times \bar{C}\bar{S}\bar{R} = \bar{C}\bar{S}R; \tag{A1.5-2}$$

$$CS\bar{R}\times \bar{C}SR = \bar{C}S\bar{R}; \tag{A1.5-3}$$

$$CS\bar{R}\otimes C\bar{S}R = C\bar{S}\bar{R}; \tag{A1.5-4}$$

$$CS\bar{R}\otimes CS\bar{R} = CSR; \tag{A1.5-5}$$

$$CS\bar{R}\times \bar{C}\bar{S}R = \bar{C}\bar{S}\bar{R}; \tag{A1.5-6}$$

$$CS\bar{R}\times \bar{C}S\bar{R} = \bar{C}SR; \tag{A1.5-7}$$

$$CS\bar{R}\times C\bar{S}\bar{R} = C\bar{S}R. \tag{A1.5-8}$$

- For matrix $\boxed{\bar{C}\bar{S}R}$ multiplied in turns with the others, we get, respectively:

$$\bar{C}\bar{S}R\times CSR = \bar{C}\bar{S}R; \tag{A1.6-1}$$

$$\bar{C}\bar{S}R\times \bar{C}\bar{S}\bar{R} = CS\bar{R}; \tag{A1.6-2}$$

$$\bar{C}\bar{S}R\times \bar{C}SR = C\bar{S}R; \tag{A1.6-3}$$

$$\bar{C}\bar{S}R\times C\bar{S}R = \bar{C}SR; \tag{A1.6-4}$$

$$\bar{C}\bar{S}R\times CS\bar{R} = \bar{C}\bar{S}\bar{R}; \tag{A1.6-5}$$

$$\bar{C}\bar{S}R\times \bar{C}\bar{S}R = CSR; \tag{A1.6-6}$$

$$\bar{C}\bar{S}R\times \bar{C}S\bar{R} = C\bar{S}\bar{R}; \tag{A1.6-7}$$

$$\bar{C}\bar{S}R\times C\bar{S}\bar{R} = \bar{C}S\bar{R}. \tag{A1.6-8}$$

- For matrix $\boxed{\bar{C}S\bar{R}}$ multiplied in turns with the others, we get, respectively:

$$\bar{C}S\bar{R}\times CSR = \bar{C}S\bar{R}; \tag{A1.7-1}$$

$$\bar{C}S\bar{R}\times \bar{C}\bar{S}\bar{R} = C\bar{S}R; \tag{A1.7-2}$$

$$\bar{C}S\bar{R}\times \bar{C}SR = CS\bar{R}; \tag{A1.7-3}$$

$$\bar{C}S\bar{R}\times C\bar{S}R = \bar{C}\bar{S}\bar{R}; \tag{A1.7-4}$$

$$\bar{C}S\bar{R}\times CS\bar{R} = \bar{C}SR; \tag{A1.7-5}$$

$$\bar{C}S\bar{R}\times\bar{C}\bar{S}R=C\bar{S}\bar{R}; \qquad \text{(A1.7-6)}$$

$$\bar{C}S\bar{R}\times\bar{C}S\bar{R}=CSR; \qquad \text{(A1.7-7)}$$

$$\bar{C}S\bar{R}\times C\bar{S}\bar{R}=\bar{C}\bar{S}R. \qquad \text{(A1.7-8)}$$

- For matrix $\boxed{C\bar{S}\bar{R}}$ multiplied in turns with the others, we get, respectively:

$$C\bar{S}\bar{R}\times CSR=C\bar{S}\bar{R}; \qquad \text{(A1.8-1)}$$

$$C\bar{S}\bar{R}\times\bar{C}\bar{S}\bar{R}=\bar{C}SR; \qquad \text{(A1.8-2)}$$

$$C\bar{S}\bar{R}\times\bar{C}SR=\bar{C}\bar{S}\bar{R}; \qquad \text{(A1.8-3)}$$

$$C\bar{S}\bar{R}\times C\bar{S}R=CS\bar{R}; \qquad \text{(A1.8-4)}$$

$$C\bar{S}\bar{R}\times CS\bar{R}=C\bar{S}R; \qquad \text{(A1.8-5)}$$

$$C\bar{S}\bar{R}\times\bar{C}\bar{S}R=\bar{C}S\bar{R}; \qquad \text{(A1.8-6)}$$

$$C\bar{S}\bar{R}\times\bar{C}S\bar{R}=\bar{C}\bar{S}R; \qquad \text{(A1.8-7)}$$

$$C\bar{S}\bar{R}\times C\bar{S}\bar{R}=CSR. \qquad \text{(A1.8-8)}$$

It results that indeed, in any combination, the strategic matrixes (1)–(8) form a closed point system, self-consistent; furthermore, for every set of strategic couples, we reobtain all the other points as result of crossed couplings! We will thus retain the results of strategic couples (A1.1-1)–(A1.8-8) as building an algebraic group, susceptible to bring valuable plus value to the strategic management in general, through the strategic cube and the specific strategic operations (of symmetry, too) in special, see Annex 2.2 and the Cognitive Analysis section.

ANNEX 2.2 SYMMETRY ELEMENTS OF THE CUBE

Symmetry elements of the cube are illustrated in Figure A2.1 as follows (Putz, 2016):

- An inversion center (i);
- Nine reflection or mirroring plans (*m*, from "mirror");
- Three rotation axes of order 4 (A^4) through the centers of the opposite sides, ensuring symmetry at 90° rotation;

- Six rotation axes of order 2 (A^2) through the edges of the opposite sides, ensuring symmetry at 180° rotation;
- Four rotation axes of order 3 (A^3) through the centers of the opposite sides, ensuring symmetry at 120° rotation.

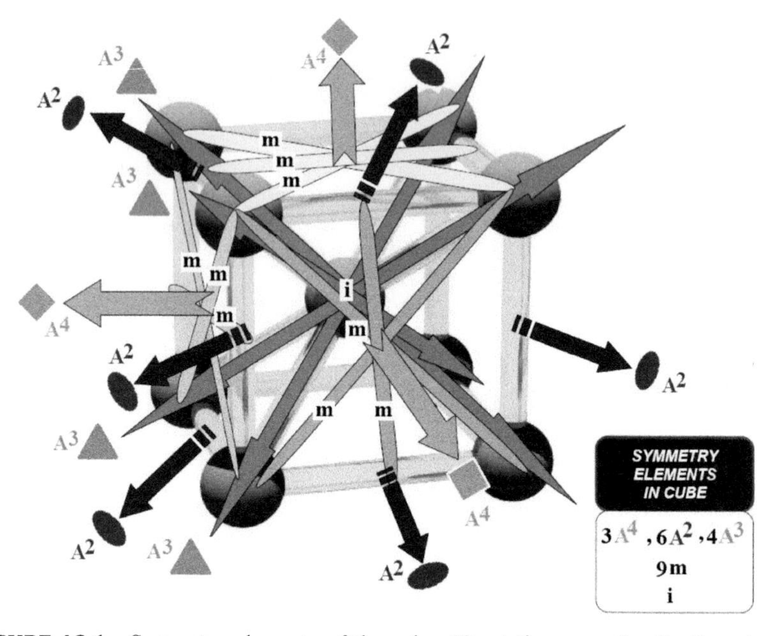

FIGURE A2.1 Symmetry elements of the cube; 13 rotation axes, 9 reflection plans, 1 inversion center.

KEYWORDS

- **competitive advantage**
- **sustainable advantage**
- **regenerative advantage**
- **cubic algebraic transformations**
- **cubic symmetries**
- **Blue Ocean strategy**
- **Red Ocean strategy**

REFERENCES

Akerlof, G. A.; Shiller, R. J. *Animal Spirits: How Human Psychology Drives the Economy and Why It Matters for Global Capitalism*; Princeton University Press: Princeton, 2010.

Alvesson, M. *Understanding Organizational Culture*; Sage: London, 2002.

Bibu, N. A.; Predişcan, M.; Sala, D. C. *Organization Management* [originally in Romanian as: *Managementul Organizaţiilor*]; Mirton Publishing House: Timişoara, Romania, 2008.

Cockburn, I. M.; Henderson, R. M.; Stern, S. Untangling the Origins of Competitive Advantage. *Strategic Manage. J.* **2000,** *21*, 1123–1145.

David, F. R. *Strategic Management: Concepts and Cases*, 13th ed.; Prentice Hall: Upper Saddle River, NJ, 2011.

Derrida, J. *Ulise Gramofon. Două cuvinte pentru Joyce*; Allfa: Bucureşti, 2000a.

Drucker, P. F. *The Practice of Management*; Harper Business: New York, 1954.

Fayol,. Administration Industrielle et Generale. *Bull. Soc. l'IndustrieMinerale* **1916,** *10* (3), 5–162.

Hodstede, G.; Hofstede, G. J.; Minkov, M. *Cultures and Organizations. Software of the Mind. Intercultural Cooperation and Its Importance for Survival*; McGaw-Hill: New York, 2010.

James, J. *The Music of the Spheres: Music, Science and the Natural Order of the Universe*. Springer Verlag: New York, 1995.

Kim, W. C.; Mauborgne, R. Blue Ocean Leadership. *Harv. Bus. Rev*. **2014,** 1–19. https://hbr.org/2014/05/blue-ocean-leadership.

Kim, W. C.; Mauborgne, R. Red Ocean Traps. *Harv. Bus. Rev.* **2015,** 1–15. https://hbr.org/2015/03/red-ocean-traps.

Lorenz, K. *Die acht Todsünden der zivilisierten Menschhei;* Piper Verlag GmbH: München, 1973.

Petrişor, I. I. *The Strategic Management. The Potentiological Approach* [originally in Romanian as: *Management Strategic. Abordare potenţiologică*]; Brumar Publishing House: Timişoara, Romania, 2007.

Porter, M. *On Competition;* Harvard Business Review Press: Brighton, 2008.

Putz, M. V.*Quantum Nanochemistry. A Fully Integrated Approach. Vol IV. Quantum Solids and Orderability*; Apple Academic Press & CRC Press: Toronto-New Jersey, 2016. http://www.appleacademicpress.com/title.php?id=9781771881364.

Putz, M. V. *Strategic Cube of the Organization Competitive Advantage.* [Originally in Romanian: *Cubul strategic al avantajului competitiv al organizaţiilor*]. MBA Thesis, Faculty of Economy Science and Business Administration, West University of Timişoara, 2017.

Putz, M. V. *The Strategic Cube of the Distinctive Advantage. Networks with Catastrophic Surfaces*, This Monograph, Chapter 3; 2019.

Stanley, T. *History of Philosophy*; Humphrey Moseley and Thomas Dring: London, 1656, 1660, 1662.

Taylor, F. W. *Shop Management*; Harper and Brothers: New York, London, 1903.

CHAPTER 3

The Strategic Cube of the Distinctive Advantage: Networks with Catastrophic Surfaces

ABSTRACT

In the postmodern approach context of a phenomenology extended at semantic, logic-fragmentary, and the managerial levels, that is, with the orchestral meaning more likely than the controller of the strategic management, arises the possibility of quasi-predictable dynamic modeling (quasi-stable), yet also unpredictable (instable) business systems, by the dynamic theory of catastrophes; remarkably, this onduliform dynamic context allows both the chaos modeling (on strategic behavior domains between an attractor and a repeller), but also the controlled change in action plans through orchestration at the network levels based on topological potential obtained by solving the extreme equations of the polynomial of associated catastrophes surfaces. The overlapping of the strategic cube of the competitive advantage (with the eighth forms of traditional business) with the strategic cube of the network orchestrated competition, for each type of dynamic on the catastrophe surface, allows identification of the strategic action plan and the dominant tendency from the traditional business perspective manifested in the network. With such analysis, the "animal spirits" behavior shows naturally as being excluded from the network orchestration—this way revealing the strategic cube as naturally selecting against the nondistinctive advantage intrinsic value (competitive, sustainable, and regenerative).

Motto:

"It's better to get to the end of the right road by oscillating
Than walking firmly on the wrong road!"

—Saint Augustine, Confessions

3.1 INTRODUCTION

Modernism! Postmodernism!

From reality to the emotion of reality!

Postmodernist fiction (McHale, 1987):

- Change of domination: hetero cosmos
- Between author and God
- Social construction of reality
- Possible worlds. Worlds in collision. Parallel lines, hesitation, triviality, resistance. Worlds being cancelled. Sense (absence) of an ending. Revisited hesitation.
- Included–excluded third parties and crossroads.
- In the area. Toward endless regress. Which roll?
- How to build a zone. There is something. Something happened. Hypertrophy.
- Intertextual zones. The fantastic dislocated. Characters in search of an author. Fictions in the abyss.
- Cat litter, litanies, phrases with broken spine. A spatial dislocation of words. The schizoid text. Authority and short circuit.
- How I have learnt not to worry and love postmodernism—love and death!

Everyday things' destiny (Norman, 2002):

- "Design can be the upper limit of the competitive advantage" (Tom Peters)
- "Button pushers throughout the world, unite!" (Los Angeles Times)
- "I'm only interested in those challenges where I believe I have something original to say—I want to be one step ahead. That's why I am now interested of the 21st-century's design" (Donald A. Norman)
- Everyday things' pathology
- Everyday actions' psychology
- Knowledge of the mind and of the world
- Knowing what to do
- It is only human to make mistakes
- Challenge of design
- User-centered design

3.2 COGNITIVE ANALYSIS

The world is curved (Smick, 2008):

- The end of the world?
- An ocean of dangerous money
- Tony Soprano is riding the Chinese dragon
- Japanese housewives occupy command positions
- The incredible compression of central banks
- The class struggle and globalization policies
- To survive or to thrive in this age of volatility?
- Nothing stays the same!

Nay!

The world is flat (Fung et al., 2008):

- Globalization 1.0: Columbus's flat world
- Globalization 2.0: Global economy and multinational companies (1800–2000)
- Globalization 3.0: Flattened world from small to tiny [according to Thomas Friedman, characterized by workflow (WWW), outsourcing, offshoring, uploading, open source, insourcing (integrated logistics)], wireless (accelerating steroids)
- "In a flat world, orchestration is one of the most important management skills. The skill of the orchestrator consists in stimulating the talent and creativity of the network, coordinating individual elements and ensuring the success of the entire process."
- What does it mean <country of origin> or <commercial bilateral deficit> in the contemporary world?
- "Imagine an eBay where the existing products are not only bought and sold, but are also created by the network!"
- "When you own your production amenities, your main concern becomes, up to a point, the use of your own capacities instead of satisfying the customers' requests. The network orchestrator identifies a need of the customer and then builds a delivery chain for manufacturing, or a service value chain, in order to satisfy this need."
- "The tighter the chain around the orchestration instead of property, the more flexible it will get, and the orchestrating skills will multiply."

- "The necessity to orchestrate the network becomes even more obvious when it does not exist or as a consequence of failures through relocation operations, externalization, or strategic alliances."
- Bumps, mountains, and highways—the need for balance! New roads of silk!
- While mountain chains evolved along the centuries, the new "political mountains" and "tunnels" of the modern world can change in months or even days.
- "The orchestrator focuses on achieving the optimal processes in a global network in order to deliver the right product to the right place, at the right time and the best price."
- "On a team race, the helmsman is not rowing. The helmsman is orchestrating!"
- "Competition means more than just a company against another company; it is rather a supply chain against another supply chain. This is a different vision on partnership and a wider outlook on the company itself."
- "Competition between networks means that the company with access to the best chains will exceed today its rivals, but will have the capacity to exceed them in the future too. The situation is similar to the force represented by the backup players of a sports team."
- "Networks can work smarter than the individuals or companies they consist of. Orchestration is the element that ensures intelligence to a smart network."
- "Although information technology is the pivot flattening the world, the nerve center of any network or enterprise is composed of human judgment, trust, human relations, and business processes."
- "Each node of the network has connections with other networks. The networks are connected to other networks. That means that the shock waves coming from a network may affect the entire system."
- "To integrate acquisitions effectively, companies do not only need technology. They must also create plug-and-play cultures, as well."
- "The network orchestrator must design these features so that they support a vision of the network," viz., back-office (e.g., highly standardized accountancy) versus middle-office (flexibility) versus front-office (customization delivery and compliance of the vendor)."
- "A modular and determinable approach of the business supports rapid and sustainable growth."

- Where's the stick? Without penalties in case of failure.
- "The road through this flat world begins with customer's needs and ends with solutions proposed by the customer" ... linearization of demands!"
- "Price reductions are considered a deficiency in the manufacturing process."
- "In a flat world with unpredictable demand is more difficult and important to be able to avoid price reductions and running out of stocks."
- "Rethinking the supply chain and transforming the customer from a passive receptor into an active participant, companies have the opportunity to recover the difference between the factory and the sale price."
- "By supply companies have sufficient information about customers, about the legal environment and potential competition."
- The market follows the factory!
- "The vertically integrated company began to disintegrate. To stop it evolving toward an unpleasant association of elements, an orchestrator must stand in front!"
- "In a flat world, there are many opportunities to apply the defining elements for network orchestration within the business, beyond the production stage."
- "The strength of a company does not necessarily stand in the competences it holds, but rather the skills that it can connect to. That means its potential to connect to other competences—orchestration potential of a network—and the learning ability may become as important as company-specific skills."
- "The network orchestration requires a certain degree of humility and flexibility of the leaders, but also openness to diversity, attention to environmental changes, and integrity."

3.3 METHODOLOGICAL ANALYSIS

We start from The Theory of Catastrophes (Thom, 1973; Woodcock and Poston, 1974; Zeeman, 1977; Zahler and Sussman, 1977; Guckenheimer, 1978; Woodcock and Davis, 1978; Saunders, 1980; Thompson, 1982; Arnold, 1992; Silvi and Savin, 1994; Sanns, 2000; Putz, 2006); this:

- Allows study of critical states and singularities
- Gives a physical pattern that allows understanding the transition phenomena
- Highlights the importance of control parameters in the evolution of natural systems

It is worth mentioning that the parent of The Theory of Catastrophes, Rene Thom:

- In 1972 publishes "Structural stability and morphogenesis"
- Proposes the modeling of discontinuous changes occurring in natural phenomena by using dynamical systems theory
- Highlights the importance of structural stability
- Describes the natural system by one of the seven elementary local catastrophes:
    - Fold catastrophe
    - Catastrophe of return
    - Dovetail catastrophe
    - Butterfly catastrophe
    - Elliptical umbilical catastrophe
    - Hyperbolic umbilical catastrophe
    - Parabolic umbilical catastrophe

This theory points attractor = local minimum points (Fig. 3.1) with specific properties:

- Two cycles that attract each other are two possible minimum states of the system
- The cycles "communicate" through global interaction (via an accretion phenomenon)

The mathematical theory of catastrophes of Thom describes how for a system—with a continuous action on a so-called control space C^k parameterized by coefficients c_k generates a sudden change in a so-called space of behavior I^m described by variables x_m, these changes being marked by the occurrence of singularities in the combined space (synergistic)

$$\eta(c_k, x_m): C^k \times I^m \rightarrow \Re \tag{3.1}$$

generating the so-called generic potentials η (c_k, x_m) for the system. Thus, catastrophes consist of a set of critical points (c_k, x_m) for which the gradient of potential is canceled:

$$M^{k \times m} = \left\{ (c_k, n_m) \in C^k \times I^m \middle| \nabla_{x_m} \eta(c_k, n_m) = 0 \right\} \quad (3.2)$$

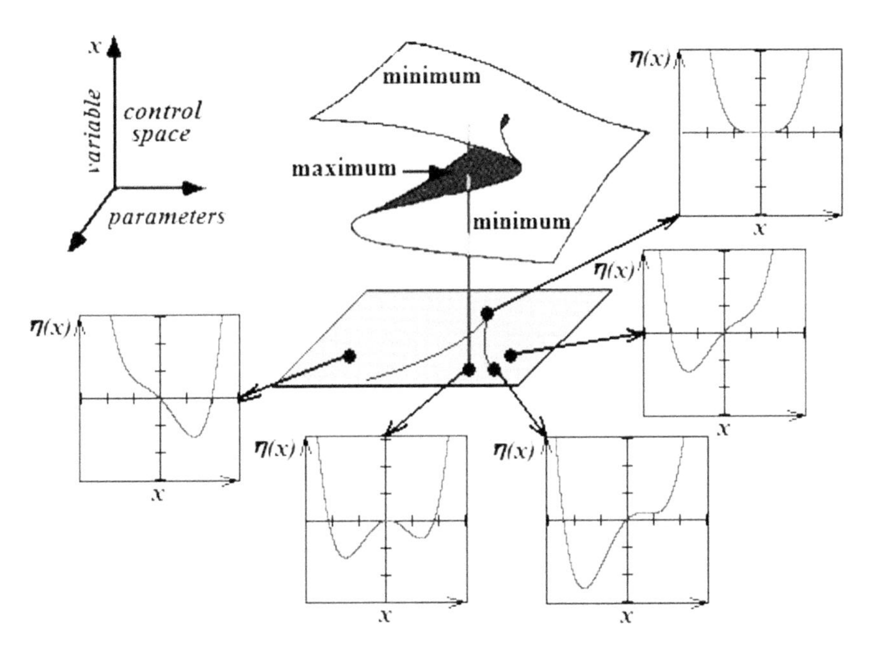

FIGURE 3.1 Representations in multiple projection of a catastrophe surface (return one here), highlighting the cycles identifying cycles focused on metastable minimums and the interaction between them.

Thus, it brings us to an even more accurate definition: a catastrophe is a singularity on the map $M^{k \times m} \rightarrow C^k$. From here, depending on the number of parameters in space C^k (also called codimension, k) and on the number of variables in the space I^m (also called corank, m) René Thom succeeded in classifying the potential generic catastrophes (or catastrophe surfaces/ maps); seven elementary (in the sense of their universality) catastrophes resulted by combining the multivariables (corank) up to dimension 2 with multiparameters (up to codimension 4); their generic polynomials forms are displayed in Table 3.1.

TABLE 3.1 Thom's Elementary Catastrophes Classification and the Correspondences of the Dominant Tendencies in the Strategic Cube of Distinctive Advantage.

Strategic type and tendency	Codimension	Corank	Thom universal polynomial
Fold	1	1	$x^3 + ux$
Quasi-stability on the level of value/creation and goal—see Fig. 3.2, with slide to smart business (from the kleptocratic one) on the level of network orchestration			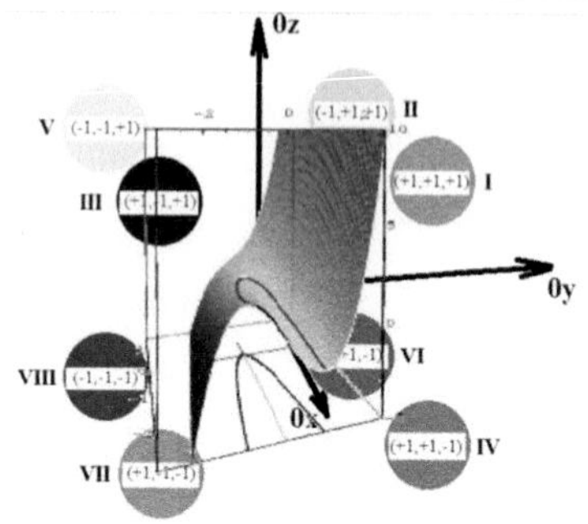
Cusp:	2	1	$x^4 + ux^2 + vx$
Quasi-stability on the level of value/ creation and goal—see Fig. 3.2, with slide to smart business from outsourcing) on the level of network orchestration			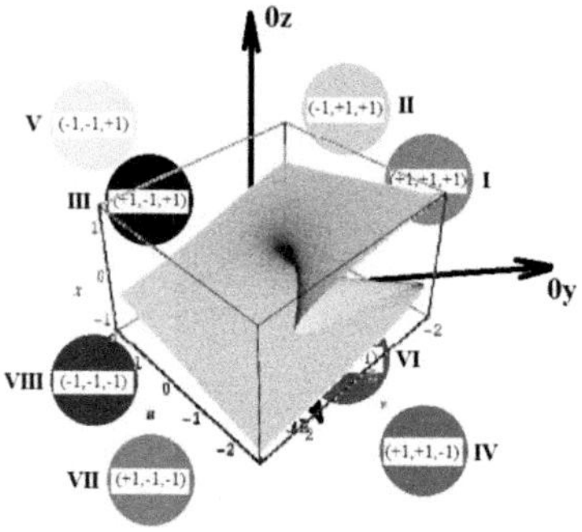
Swallowtail	3	1	$x^5 + ux^3 + vx^2 + wx$
Quasi-stability on the level of value/creation and goal but with a lot of responsibility delegation—see Fig. 3.2, with slide to polemocracy (from wise business on the level of network orchestration			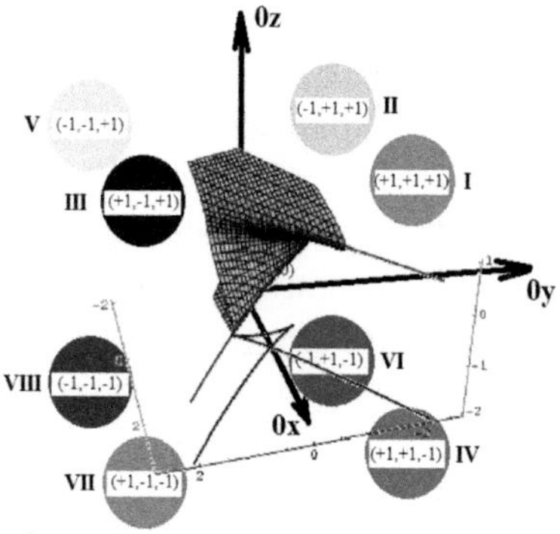

TABLE 3.1 *(Continued)*

Strategic type and tendency	Codimension	Corank	Thom universal polynomial
Hyperbolic umbilic	3	2	$x^3 + y^3 + uxy + vx + wy$
Instability with tendencies toward perverse spirits (from smart business) on the level of network orchestration			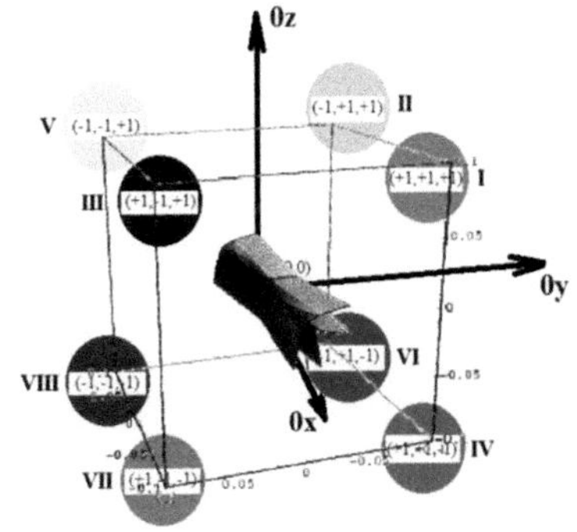
Elliptical umbilic	3	2	$x^3 + xy^2 + u(x^2 + y^2) + vx + wy$
stability in delegating responsibilities and competences in the network, see Fig. 3.2, with oscillations of integration–disintegration (between polemocracy and wise business)_on the level of network orchestration			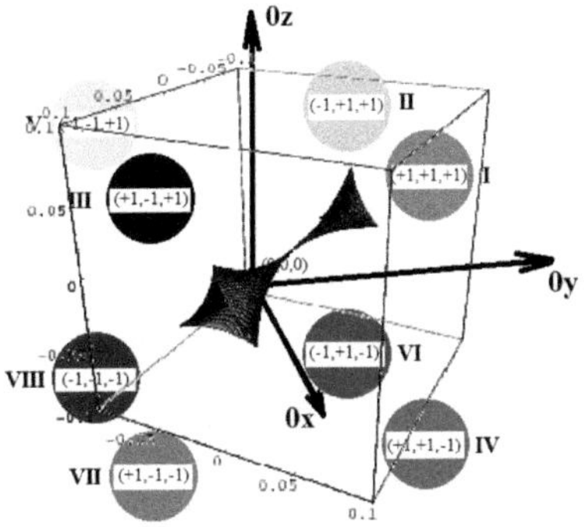
Butterfly	4	1	$x^6 + ux^4 + vx^3 + wx^2 + tx$
Quasi-stability on the level of value/creation and goal–see Fig. 3.2, with shift to smart business (from the wise one) by diminishing delegation of competences on the level of network orchestration			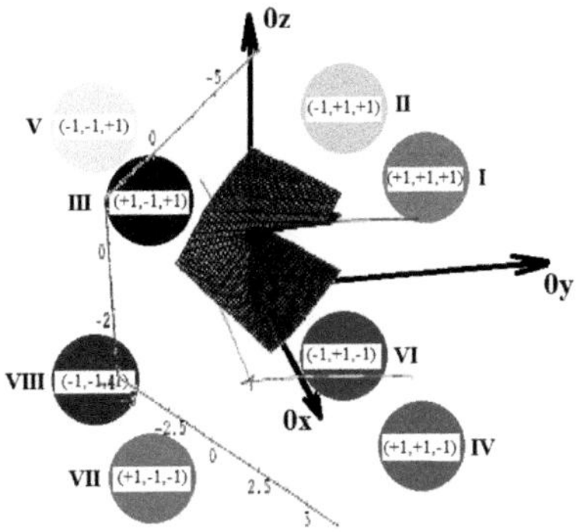

TABLE 3.1 *(Continued)*

Strategic type and tendency	Codimension	Corank	Thom universal polynomial
Parabolic umbilic	4	2	$x^2y + y^4 + ux^2 + vy^2 + wx + ty$
instability on the level of value and goal, see Fig. 3.2 due to too much delegation of responsibilities and competences, with outsourcing domination (combined with polemocracy and smart business) on the level of network			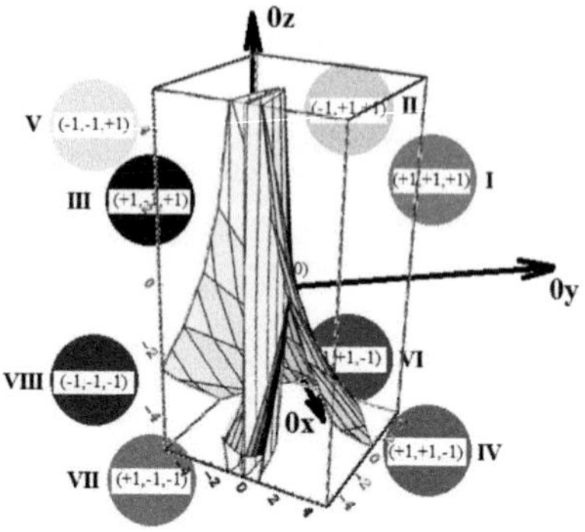

To be noticed that for superior rank derivates of generic potentials it will be stated that for those control parameters c_k* for which the Laplacian generic potential is canceled:

$$\Delta_x \eta\ (c_k{}^*, x_m) = 0 \tag{3.3}$$

the so-called bifurcation points are obtained. On the other hand, the control parameters c* for which the Laplacian in a critical point is not canceled defines the stability domain of the respective critical point. It is immediate then that for small disturbances of functionsη (c*, n), the system shifts from the domain of stability into another; if the opposite happens, it is said that the system is located into a domain of structural stability.

Furthermore, the cases described above correspond to the balance limit of the general evolution of natural systems:

$$f\left(c_k;t;\eta(c_k;x_m);\frac{\partial\eta(c_k;x_m)}{\partial t}\right)=0 \tag{3.4}$$

where now the behavior space is further on parameterized through temporal paths x_m (c_k,t).

Connection to balance is regained through the stationary condition, imposed at the critical points. Thus, a critical point in the stationary mode $t \to +\infty$ (called as the ω-limit) corresponds to an attractor and forms the basin corresponding to it, while the stationary mode for which for the $t \to -\infty$ (called as the α-limit) discloses a repeat.

According to postmodernist strategic management, the cube of the distinctive advantage (Putz, 2017, 2019) can be correlated with the cubic diagram of competitors in the orchestrated network of Figure 3.2. On this paradigm we can notice:

- Focus on integration to the network compared with specialization in the traditional business;
- Focus on network compared with business in achieving the goals (of the business and of the network);
- Control turns into delegation in the management practice;
- The catastrophe surfaces corresponds with the distinctive behavioral business states of the strategic cube; accordingly, one can build a chain of catastrophic waves in the cube of the competitive advantage, allowing a temporal unfolding and networking as depending on the advanced or retarded states respecting the referential "wise business"—the ideal state of meta-economy. However, all business states can be reached by catastrophic waves, except for the state of the perverse spirits which is naturally excluded from this dual construction of the cube of the distinctive advantage in the network orchestrated competition. Consequently, there are left only seven states and behaviors of companies in the network, and they associate further on, in a dynamic environment and having a dynamic behavior, in their turn, shaped by the seven Thom, elementary catastrophes, as described and discussed in Table 3.1.
- Control of catastrophe surfaces through topological potentials (Petrişor, 2007).

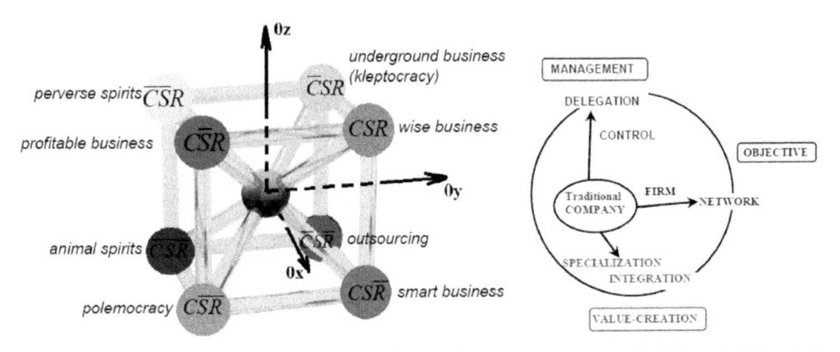

FIGURE 3.2 Strategic cube of distinctive advantage (Putz, 2017, 2019)—left, and directions of strategic development from the traditional company to network, on the components 3D space, in the cognitive cube of strategic management orchestrated in the network (Fung et al., 2008)—right.

The analysis of Table 3.1 is inherently dynamic, combines tendencies and constructive and destructive interfaces in the network through sudden modifications/changes, synergic, on the level of businesses integrated in a network, orchestrated mainly with focus on creating value through learning.

3.4 CONCLUSIONS AND PERSPECTIVES

The paradox can start with the test of Botton (2004):

Select the antonyms in these four words:			
stubborn	fake	flexible	obscure
	and		
Say which word has the same meaning with the first one:			
inexpugnable	bizarre	emptiness	
nominal	excessive	didactic	

and may continue recursively in the postmodern manner (free creative thinking) with:

- Umberto Eco: Infinite list, glocal
- Solomon Marcus's postmodernism with universal paradigms
- Postmodern economy with anti-economy or economy in the era of chaos, of quantum bifurcations, and fuzzy decisions—laying foundation to new hypotheses for eco–glocal–economy!
- Neuronal network—as reply to quantum connections (a.k.a chemical connections in molecules), as states of balance and anti-balance Nash....generating a kind of glocal matrix.

The paradox is not so paradoxical when viewed on the whole, from the dynamic, inclusive perspective, where everything interacts with everything, when understood attractions and rejections occur, chaos and bifurcations learned, attractors and strategic repellers. The catastrophe theory effectively captures this wave characteristic of business states and their interferences, while unifying them by the nonredundant reduction to seven dynamic paradigms, with a dispersive time–behavior; they can properly be phenomenological analyzed here, and algorithmically analyzed in other dedicated studies (Putz, 2017, 2019), as based on the correspondence between the strategic cube of distinct advantage and the cube of the competitor paradigm orchestrated in and through a generic network.

KEYWORDS

- **postmodernism**
- **strategic cube of distinctive advantage**
- **network orchestrated competition cube**
- **topological potentials**
- **universal catastrophes**

REFERENCES

Arnold, V. I. *Catastrophe Theory*, 3rd ed.; Springer-Verlag: Berlin, 1992.

de Botton, A. *Status Anxiety;* [2015: read in Romanian language under the title *Statut şi Anxietate*. Vellant: Bucureşti], 2004.

Fung, V. K.; Fung, W. K.; Wind, Y. *Competing in a Flat World: Building Enterprises for Borderless World*, 1st ed.; Pearson Education: New Jersey, 2008.

Guckenheimer, J. The Catastrophe Controversy. *Math. Intell.* **1978,** *1*, 15–20.

McHale, B. *Postmodernist Fiction*; Routledge: London & New York, 1987.

Norman, D. A. *The Design of Everyday Things* [1988: *The Psychology of Everyday Things*]; Basic Books: New York, 2002.

Putz, M. V. The Deviance, the Minors, and the Catastrophe Theory [originally in Italian as: *Devianza, minori e teoria della catastrofe*]. In: *The Existential Catastrophes: The Anatomy of the Minors' Disadvantage*[originally in Italian as: *Catastrofi esistenziali. Anatomia del disagio giovanile*]; Palazzo, S., Ed.; Periferia Publishing House: Cosenza, Italy, 2006.

Petrişor, I. I. *The Strategic Management: The Potentiological Approach* [originally in Romanian as: *Management Strategic. Abordare potenţiologică*]; Brumar Publishing House: Timişoara, Romania, 2007.

Putz, M. V. Strategic Cube of the Organization Competitive Advantage. [originally in Romanian: *Cubul strategic al avantajului competitiv al organizaţiilor*]. MBA Thesis, Faculty of Economy Science and Business Administration, West University of Timişoara, 2017.

Putz, M. V. *The Strategic Cube of the Distinctive Advantage: Epistemological Approach*, 2019 (Chapter 2 of the present monograph).

Sanns, W. *Catastrophe Theory with Mathematica: A Geometric Approach;* Germany: DAV, 2000.

Saunders, P. T. *An Introduction to Catastrophe Theory;* Cambridge University Press, Cambridge, England, 1980.

Silvi, B.; Savin, A. Classification of Chemical Bonds Based on Topological Analysis of Electron Localization Functions. *Nature* **1994,** *371*, 683–686.

Smick, D. M. *The World is Curved: Hidden Dangers to the Global Economy*; Peguin Group: New York, 2008.

Thom, R. *Stabilitè Structurelle et Morphogènése*; Benjamin-Addison-Wesley: New York, 1973.

Thompson, J. M. T. *Instabilities and Catastrophes in Science and Engineering;* Wiley: New York, 1982.

Woodcock, A. E. R.; Poston, T. A. *A Geometrical Study of the Elementary Catastrophes*; Springer-Verlag: Berlin, 1974.

Woodcock, A. E. R.; Davis, M. *Catastrophe Theory;* E. P. Dutton: New York, 1978.

Zahler, R.; Sussman, H. J. Claims and Accomplishments of Applied Catastrophe Theory. *Nature* **1977,** *269*, 759–763.

Zeeman, E. C. *Catastrophe Theory–Selected Papers 1972–1977;* Addison-Wesley: Reading, MA.

CHAPTER 4

Strategic Innovating Paths for the Distinctive Advantage: The Changing Management Faraway from Equilibrium

ABSTRACT

Innovation, as a process, is applied to the strategic competitive, sustainable, and regeneration advantage, by a min–max modeling of probabilities and strategic entropy in the so-called strategic cube of distinctive advantage and the Hamming–Putz algebra (space). Thus, an innovation hierarchy centered in the trifecta point (three times perfect) of the competitive, sustainable, and regeneration advantage is obtained. It is presented as the referential of zero entropy (maximum complexity) followed by the "animal spirits" strategy (specific to red ocean strategy) and two groups of strategic approach with successively increasing the entropy (less and less utility) of business; each group consists of three forms of distinctive advantage with equal entropies. The results show, at the same time, universality in the strategic modeling by path maximization (options) between different types of distinctive advantages. Their continuous change along a Hamiltonian chain (by which each strategy is touched one time on a strategic chain) is promoted, thus supporting the paradigm change from the productivity economy toward utility economy (or wise business) with the aid of the postmodern strategic innovation.

4.1 INTRODUCTION

The modern theory of entropic value of products and economic services is currently reconsidered in the postmodern version of the path (chain) of maximum value of innovation in the context of globalizing competitiveness, by greening productivity and social utility of the economical act in general and of the distinct advantage in particular (Porter, 2008; Petrisor, 2007). Thus, we distinguish three goals for distinct advantage in business, namely (Petrisor, 2007; Norman, 2010):

- *Competitive advantage* = low cost, high return (through efficiency, innovation in business administration) = profitable business (*1G, 2G, 3G*: mechanization, electrification, digitalization, respectively)
- *Sustainable advantage* = saving resources (by effectiveness, supply chain innovation, eco-innovation) = smart business (*4G*: networking, total connection)
- *Regenerative advantage* = economy of utility, product design/ process (by implication, value chain innovation, social innovation) = wise business (*5G:* overall design, using integrated use, full usefulness).

These three properties may exist independently, generating, respectively, the so-called *business strategy group* (Putz, 2017, 2019):

- Polemocratic type (lose–lose);
- Outsourcing type (outsourcing);
- Perverse spirit type (win–lose/lose–win).

When combined, they result in the so-called *cube of distinct advantage* covering the 3D space by successive coexistence of the three properties (and the nuances of those above), that is, the competitive, sustainable, and regenerative advantages, optimally directed to the referential business (111), Figure 4.1.

However, one can shape also the so-called "algebraic order" of the strategic sequences deployed in the space of the distinct advantage by successive exchange of a characteristic (information) with its negative (e.g., competitive with noncompetitive, sustainable with unsustainable, regenerative with nonregenerative), singularly or in combination with 3D algebraic space. So we get the *algebraic canonical path* (*formally, from "plus to minus infinite" in the cubic universe of the distinctive space*):

$$
\begin{aligned}
&(1\ \ 1\ \ 1)\big|_{\substack{WISE\\BUSINESS}}\\
&\rightarrow(-1\ \ 1\ \ 1)\big|_{\substack{UNDERGROUND\\BUSINESS\\(KLEPTOCRACY)}}\rightarrow(1\ \ -1\ \ 1)\big|_{\substack{PROFITABLE\\BUSINESS}}\rightarrow(1\ \ 1\ \ -1)\big|_{\substack{SMART\\BUSINESS}}\\
&\rightarrow(-1\ \ -1\ \ 1)\big|_{\substack{PERVERSE\\SPIRITS}}\rightarrow(-1\ \ 1\ \ -1)\big|_{OUTSOURCING}\rightarrow(1\ \ -1\ \ -1)\big|_{POLEMOCRACY}\\
&\rightarrow(-1\ \ -1\ \ -1)\big|_{\substack{ANIMAL\\SPIRITS}}
\end{aligned}
\tag{4.1}
$$

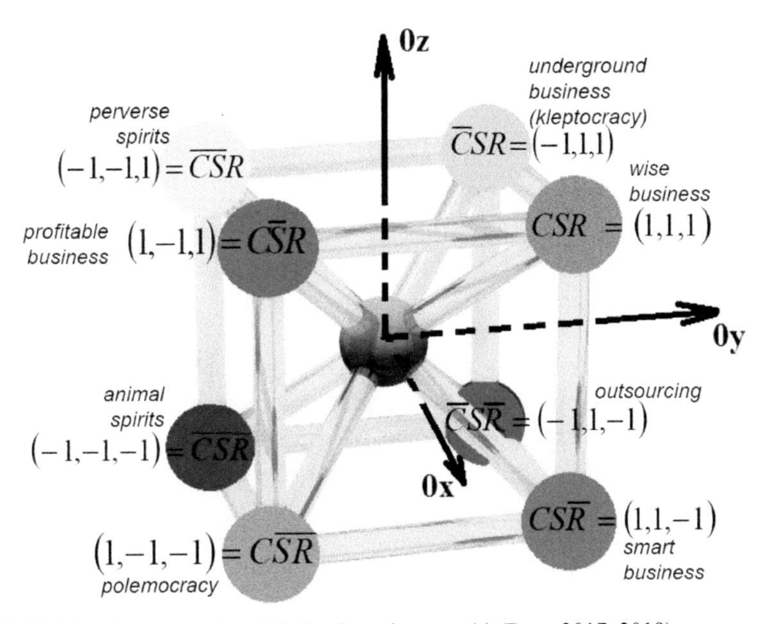

FIGURE 4.1 Strategic cube of distinctive advantage(s) (Putz, 2017, 2019).

Still the issue of the "universality" of the algebraic path (1) remains, once noticing that it appears as being discontinuous when it is "projected" onto the strategic cube of Figure 4.1; thus, the possibility of constituting/building the so-called "geometrical path" which can be chosen in several ways occurs, so "unifying" any strategic state/strategic point to the one of the wise business (111), as the final goal of total change/innovation. The present study pleads for this choice to be made accordingly to the principle of *maximal* strategic paths of innovation, as it corresponds to the maximization of energetic information (in close parallelism of physical resources a.k.a free energy, so transposed for the complex organization as the degree of innovation). In such context, one can seek to minimize the entropy along the chain of the product/process value chosen as such. Besides, one likes to strategically avoid (for sustaining the innovation strategy) the degradation of the *circulating economic object* and its innovation degree. Thus, the enhancing of a business viability and degree by (econo-ecological) utility at the expense of maximal entropy is the main purpose of such behavioral path; specifically, the manifestation of minimal paths, as stationary limit, should be avoided since unbalancing the meta-economy; and as a consequence avoiding the appearance of maximum of disorder, which equivalents with internal interference within a system (business or organization), leading to the unwished maximal entropic, a.k.a minimum of information–organization–complexity, and less strategy.

4.2 COGNITIVE ANALYSIS

Motivation and cognitive support of this study falls within the following conceptual considerations:

- *The role of entropy in economy* (Marcus, 2011): economic activity (especially production and organization) is associated with degradation of entropy (increase of information, organization) for a *closed system;* but, naturally, in a global, ecological, and social context—economic systems are *open systems* with increasing trend of entropy, which, at the limit (faraway from equilibrium) leads to "heat death" of the business space (Georgescu-Roegen, 1979), of the products while eroding the services, noticing the destruction of markets, and occurrence of financial crises on economic–socio-ecological bases (Haret, 1910); the solution would be the consumption of "negentropy" (Schödinger, 1944) from the environment, so that the system/economic act in question (product or process) to create distinct advantages through utility (Pikler, 1954);
- *From natural process optimization to economic decisions under uncertainty and maximizing social satisfaction*: the principles according to which the minimum of energy corresponds to maximum of information under equilibrium conditions may be generalized for open systems (manifesting exchange energy and information with the outside), so that couples min–max and max–min are the directories in further optimizing systems faraway from equilibrium (especially in external interaction); This behavior is confirmed

by modeling managerial decisions (Ionescu and Cazan, 2007) addressing decisions under uncertainty (in open systems) through a series of techniques combining the extremes: *max–max* (optimistic criterion), *max–min* (pessimistic criterion) *max* {sum of max and min} for the realistic criterion, maximizing the profit expectation so minimizing the loss of expectation under the criterion of maximum probability for conditioned payments, etc.; not least, *sociomatics* (Simoi, 1978) shapes values such as: *social groups* (the multi-intentional systems, social censorship of individual decisions social wealth), *social power* and relations, *social relations* (class parts and relational borders), *social co-action* (social association, gregarious tropisms, solidarity, generosity, trust and social hierarchy), *social interaction* (competition, fight, decisional constraint, domination and social ascent, social position and circumstance), together with *social information* (communication and social manipulation, schools and elites, educational strategies)—and much more—all in terms of maximizing collective, local, group, or global information in the reciprocal inter-dynamics.

- *Hamming space* {0,1}: identifying errors and correction of errors by substitution of what is known as bits in information theory (Hamming, 1950); however, in economy they can be considered as the econo-bits. It is nevertheless here extended by considering the counter information, so the counter forces specific to competition (including underground competition) act within an *extended Hamming space,* that is, within 3D space {−1,0,1}.

With these considerations, strategic paths and changes the cube of distinct advantage of Figure 4.1 will use, in a synergic manner, (conjugated) the following ideas, abstracted from the above, namely:

- *The presence of entropy* in eroding productivity and enhancing usefulness;
- *Combining principles min–max in optimizing decisions/strategic development plans;*
- *Combining information (forces) and counterinformation (counter-forces),* values of truth and their negations, in order to formulate a ecolo-economy of change through competitive, sustainable, and regenerative advantage.

4.3 METHODOLOGICAL ANALYSIS

Analysis of strategic hierarchies under conditions of distinct advantage (Fig. 4.1) involves the evaluation of the two types of strategies (with mathematical value):

- A strategic algebraic approach representing quantification, abstraction to universal econometrics, less dependent on the environment (specific to *closed economic systems,* organizations and their internal changes);
- A strategic geometric approach: the interpretation of vicinities, modeling paths, configuring business space, optimizing economic decisions (borders) are determinant factors in business decision-making, innovation and change (specific to *open economic systems,* markets for products and services in local, regional, and global dynamics).

The two approaches may seem contrary, when in fact they are only complementary, as it will be considered below.

Regarding the strategic algebraic approach, as instrumental basis the so-called *Hamming distance* (3D) will be taken: it represents the deviation from the perfect configuration, from maximal information, from the trifecta point of the competitive, sustainable, and regenerative advantage (111).

The present approach takes into consideration the extension of the Hamming space presented at the end of the previous section; this extension can correspond to the included third party—namely, the *tertium datur* (Nicolescu, 2007), by interpreting the achievement negatives on the (sign) coordinates of the trifecta point (111) as the opposed forces (in either meaning of distance, dispersion or diffusion information). It is worth mentioning the fact that within the extended Hamming space the corresponding algebra (or the so-called Hamming–Putz algebra) acts with its basic operations:

$$\{+1+1=0;\ -1+(-1)=0;\ +1+(-1)=0;\ -1+1=0;\ +1+0=+1;-1+0=-1\} \quad (4.2)$$

Thus, the *Hamming–Putz distance* can be introduced as being equal to the number of econo-bits necessary to connect any point/state of the distinct advantage (one of the strategies) to the trifecta point of the competitive, sustainable, and regenerative advantage. This can be done by successive operations as applied for each coordinate of the extended Hamming space, carrying the operations of the Hamming–Putz algebra, see eq (4.2); actually, the following formula is applied:

$$\Delta_{xyz}^{\mathrm{HP}}\left[\begin{pmatrix}x & y & z\end{pmatrix}-\begin{pmatrix}C & D & P\end{pmatrix}\right]:=\hat{\Theta}_{aCDP}\begin{pmatrix}x & y & z\end{pmatrix}\Big|\begin{pmatrix}x\mapsto 1 & y\mapsto 1 & z\mapsto 1\end{pmatrix} \quad (4.3)$$

with $\hat{\Theta}_{aCDP}\begin{pmatrix}x & y & z\end{pmatrix}\Big|\begin{pmatrix}x\mapsto 1 & y\mapsto 1 & z\mapsto 1\end{pmatrix}$ the operator of permutations who "moves" the strategic point (xyz) to the optimal strategic point (111); the cardinal of the number generated by these permutations (dimension of this number) represents the Hamming–Putz distance (HP).

Based on the definition relation (eq 4.3) the Hamming–Putz distances are customized in an algebraic manner (automated minimal), as it is summarized in Table 4.1 for each of the states of the distinctive advantage of Figure 4.1.

TABLE 4.1 Determining the Algebraic Hamming–Putz Distances for Strategies Specific to Distinct Advantage of Figure 4.1.

The algebraic minimum path, eq (4.2)	Information justification
$\min\Delta_{111}^{\mathrm{HP}}=0$	no error/no correction
$\min\Delta_{-111}^{\mathrm{HP}}=\min\Delta_{1-11}^{\mathrm{HP}}=\min\Delta_{11-1}^{\mathrm{HP}}=2$	$\left\|\left\{-1\underline{+1}=0\ \&\ 0\underline{+1}=1\right\}\right.$
$\min\Delta_{-1-11}^{\mathrm{HP}}=\min\Delta_{1-1-1}^{\mathrm{HP}}=\min\Delta_{-11-1}^{\mathrm{HP}}=4$	$\left\|\begin{Bmatrix}-1\underline{+1}=0\ \&\ 0\underline{+1}=1\\-1\underline{+1}=0\ \&\ 0\underline{+1}=1\end{Bmatrix}\right.$
$\min\Delta_{-1-1-1}^{\mathrm{HP}}=6$	$\left\|\begin{Bmatrix}-1\underline{+1}=0\ \&\ 0\underline{+1}=1\\-1\underline{+1}=0\ \&\ 0\underline{+1}=1\\-1\underline{+1}=0\ \&\ 0\underline{+1}=1\end{Bmatrix}\right.$

The maximum Hamming–Putz paths are obtained by applying the combined strategic geometry with the *ergodic principle* (i.e., filling the Hamming–Putz space) with Hamiltonian paths (which go only once through "a given point" or a given state). They start from every distinct advantage state of Figure 4.1, including from "wise business to wise business," on the longest possible path and "as parallel as possible to the algebraic path of the standard chain (1)"; for the antiparallel case, the negative contribution with sign "–" will be accounted; actually, we combine in an algebraic manner (with sign) the algebraic Hamming–Putz distances of Table 4.1 in the geometrical paths maximally superposed to the algebraic path (1), in the cube of the distinct advantage in Figure 4.1. The specific geometrical paths, the global calculation method and the results of the geometrical Hamming–Putz distances are specified in Table 4.2.

TABLE 4.2 Determining the Geometrical Hamming–Putz Distances, for Maximal Paths Specific to Strategies of Distinct Advantage of Figure 4.1, for Every Type of Strategy, Respectively.

The geometrical maximum path, Figure 4.1	Maximum specific path
Strategy I $\max\Delta_{111}^{\mathrm{HP}}$ $= \min(111) - \min(-111) - \min(-1-11) + \min(1-11)$ $- \min(1-1-1) - \min(-1-1-1) + \min(-11-1) + \min(11-1) + \min(111)$ $= 0 - 2 - 4 + 2 - 4 - 6 + 4 + 2 + 0$ $= -8$	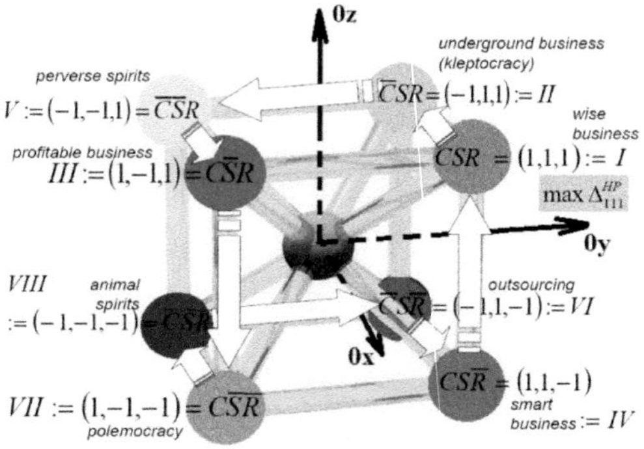 0z; underground business (kleptocracy) $\overline{C}SR=(-1,1,1):=II$; perverse spirits $V:=(-1,-1,1)=\overline{CSR}$; wise business $CSR=(1,1,1):=I$; profitable business $III:=(1,-1,1)=C\overline{S}R$; $\max\Delta_{111}^{HP}$; 0y; VIII animal spirits $:=(-1,-1,-1)=\overline{C}\overline{S}\overline{R}$; outsourcing $\overline{C}\overline{S}R=(-1,1,-1):=VI$; 0x; $CS\overline{R}=(1,1,-1)$ smart business $:=IV$; $VII:=(1,-1,-1)=C\overline{SR}$ polemocracy
Strategy II $\max\Delta_{-111}^{\mathrm{HP}}$ $= \min(-111) - \min(-1-11) + \min(1-11) - \min(1-1-1)$ $- \min(-1-1-1) + \min(-11-1) + \min(11-1) + \min(111)$ $= 2 - 4 + 2 - 4 - 6 + 4 + 2 + 0$ $= -4$	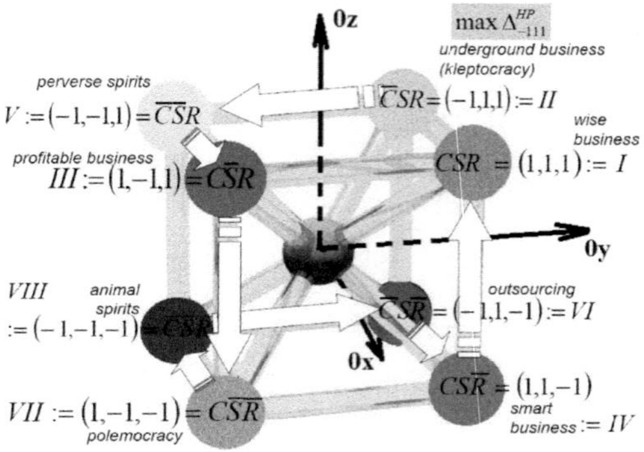 0z; $\max\Delta_{-111}^{HP}$; underground business (kleptocracy) $\overline{C}SR=(-1,1,1):=II$; perverse spirits $V:=(-1,-1,1)=\overline{CSR}$; wise business $CSR=(1,1,1):=I$; profitable business $III:=(1,-1,1)=C\overline{S}R$; 0y; VIII animal spirits $:=(-1,-1,-1)=\overline{C}\overline{S}\overline{R}$; outsourcing $\overline{C}\overline{S}R=(-1,1,-1):=VI$; 0x; $CS\overline{R}=(1,1,-1)$ smart business $:=IV$; $VII:=(1,-1,-1)=C\overline{SR}$ polemocracy

TABLE 4.2 *(Continued)*

The geometrical maximum path, Figure 4.1	Maximum specific path
Strategy III $\max \Delta^{HP}_{1-11}$ $= \min(1-11) - \min(-1-11) + \min(-111) - \min(-11-1)$ $- \min(-1-1-1) + \min(1-1-1) + \min(11-1) + \min(111)$ $= 2-4+2-4-6+4+2+0$ $= -4$	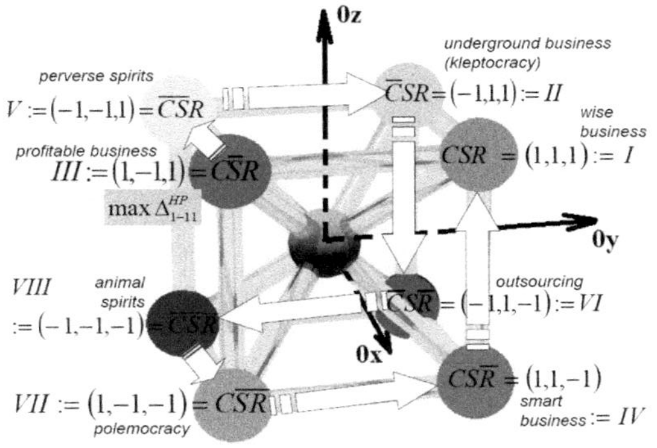
Strategy IV $\max \Delta^{HP}_{11-1}$ $= \min(11-1) - \min(-11-1) - \min(-1-1-1) + \min(1-1-1)$ $+ \min(1-11) - \min(-1-11) + \min(-111) + \min(111)$ $= 2-4-6+4+2-4+2+0$ $= -4$	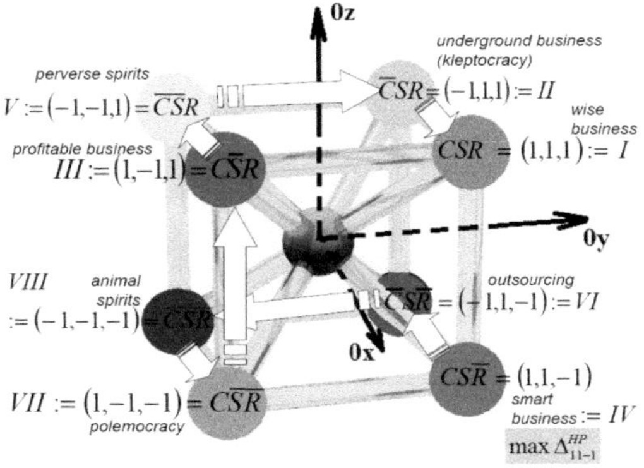

TABLE 4.2 *(Continued)*

The geometrical maximum path, Figure 4.1	Maximum specific path
Strategy V $\max\Delta^{\text{HP}}_{-1-11}$ $=\min(-1-11)+\min(1-11)-\min(1-1-1)-\min(-1-1-1)$ $+\min(-11-1)+\min(11-1)+\min(111)$ $=4+2-4-6+4+2+0$ $=2$	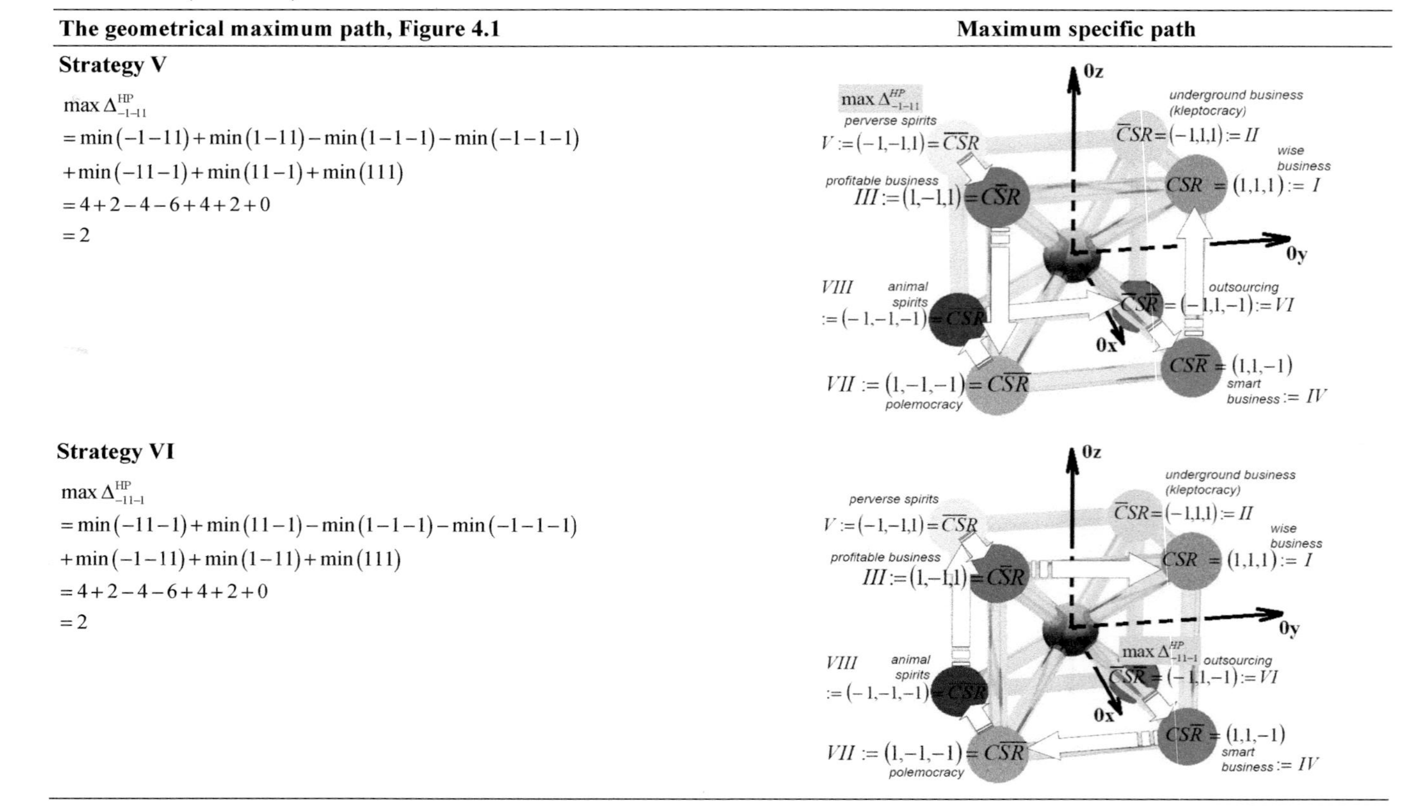
Strategy VI $\max\Delta^{\text{HP}}_{-11-1}$ $=\min(-11-1)+\min(11-1)-\min(1-1-1)-\min(-1-1-1)$ $+\min(-1-11)+\min(1-11)+\min(111)$ $=4+2-4-6+4+2+0$ $=2$	

TABLE 4.2 *(Continued)*

The geometrical maximum path, Figure 4.1	Maximum specific path
Strategy VII $\max \Delta^{HP}_{1-1-1}$ $= \min(1-1-1) - \min(-1-1-1) + \min(-11-1)$ $+ \min(-111) - \min(-1-11) + \min(1-11) + \min(111)$ $= 4-6+4+2-4+2+0$ $= 2$	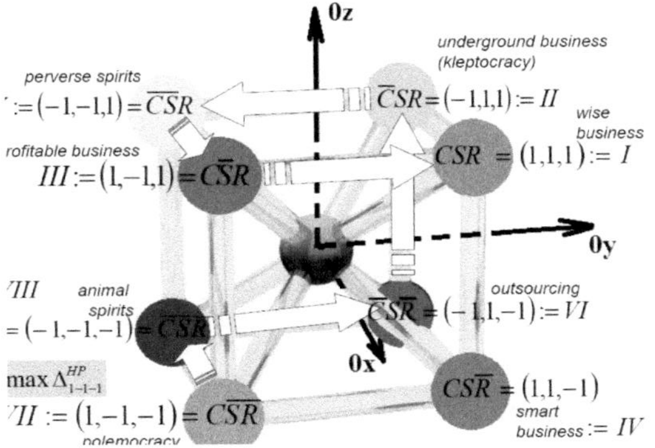 0z underground business (kleptocracy) $\overline{C}SR = (-1,1,1) := II$ perverse spirits $:= (-1,-1,1) = \overline{CSR}$ wise business $CSR = (1,1,1) := I$ rofitable business $III := (1,-1,1) = C\overline{S}R$ 0y III animal spirits $= (-1,-1,-1) = \overline{CS}R$ outsourcing $\overline{C}S\overline{R} = (-1,1,-1) := VI$ 0x $\max \Delta^{HP}_{1-1-1}$ $CS\overline{R} = (1,1,-1)$ smart business $:= IV$ $II := (1,-1,-1) = C\overline{SR}$ polemocracy
Strategy VIII $\max \Delta^{HP}_{-1-1-1}$ $= \min(-1-1-1) + \min(1-1-1) + \min(11-1) - \min(-11-1)$ $+ \min(-111) - \min(-1-11) + \min(1-11) + \min(111)$ $= 6+4+2-4+2-4+2+0$ $= 8$	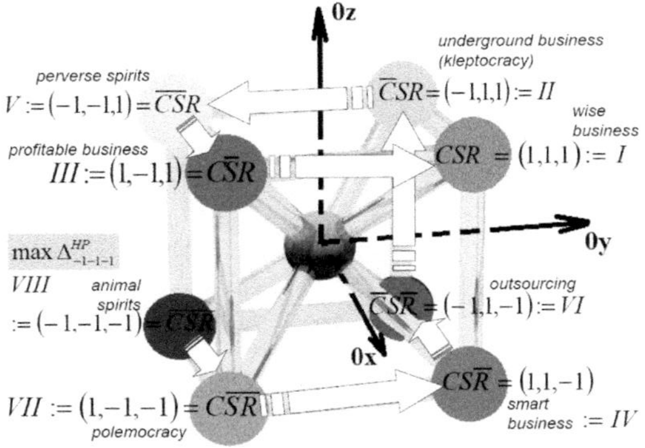 0z underground business (kleptocracy) $\overline{C}SR = (-1,1,1) := II$ perverse spirits $V := (-1,-1,1) = \overline{CSR}$ wise business $CSR = (1,1,1) := I$ profitable business $III := (1,-1,1) = C\overline{S}R$ 0y $\max \Delta^{HP}_{-1-1-1}$ VIII animal spirits $:= (-1,-1,-1) = \overline{CS}R$ outsourcing $\overline{C}S\overline{R} = (-1,1,-1) := VI$ 0x $CS\overline{R} = (1,1,-1)$ smart business $:= IV$ $VII := (1,-1,-1) = C\overline{SR}$ polemocracy

The information in Tables 4.1 and 4.2 is eventually combined to evaluate the strategic entropy $S_{\wp}$ (of the strategic path) based on the possibility of strategic transformation ($\wp$), with analytical definitions:

$$\begin{cases} S_{\wp} = \wp \ln \wp \\ \wp := 1 - \dfrac{\min \Delta_{xyz}^{HP}}{\max \Delta_{xyz}^{HP}} \end{cases} \tag{4.4}$$

and the results in Table 4.3, on strategic options of the cube of distinct advantage.

TABLE 4.3 Centralizing Triplet Strategy–Probability–Entropy for the Options of Distinct Advantage in Figure 4.1 with Results in Tables 4.1 and 4.2, by Applying the Current Formula (eq 4.4) of the Entropy of Strategic Path.

Strategy	Strategic probability	Strategic entropy
I: Wise business	$\wp(111) = 1 - \frac{0}{8} = 1$	$S_{\wp}(111) = 0$
II: Underground business (kleptocracy)	$\wp(-111) = 1 + \frac{2}{4} = 1.5$	$S_{\wp}(-111) = 0.608$
III: Profitable business	$\wp(1-11) = 1 + \frac{2}{4} = 1.5$	$S_{\wp}(1-11) = 0.608$
IV: Smart business	$\wp(11-1) = 1 + \frac{2}{4} = 1.5$	$S_{\wp}(11-1) = 0.608$
V: Perverse spirits	$\wp(-1-11) = 1 - \frac{4}{2} = -1$	$S_{\wp}(-1-11) = i\pi$
VI: Outsourcing	$\wp(1-1-1) = 1 - \frac{4}{2} = -1$	$S_{\wp}(-11-1) = i\pi$
VII: Polemocracy	$\wp(1-1-1) = 1 - \frac{4}{2} = -1$	$S_{\wp}(1-1-1) = i\pi$
VIII: Animal spirits	$\wp(-1-1-1) = 1 - \frac{6}{8} = 0.75$	$S_{\wp}(-1-1-1) = -0.216$

The results are extremely interesting, finding the following:

- *Wise business*, corresponding to trifecta point (!!!) and, respectively, to the strategy of the competitive, sustainable, and regenerative advantage (= profitable business, with saving resources, including ecological ones and with social/public utility) corresponds to zero entropy (perfect dynamic system with no erosion and dispersion, but lasting);

- "*Arrogant strategic group*" (simple negative) which consists of underground businesses (Strategy II), profitable (Strategy III), and smart ones (Strategy IV), but all having the same strategic entropy. This suggests, paradoxically, a similar utility at global level (competitive and noncompetitive) and on the social impact (immediate and regenerative);
- "*Opportunist/warlike*" *strategic group (double negative)* brings together perverse business (Strategy V), of outsourcing (Strategy VI) and polemocratic (Strategy VII), again with equal strategic entropy, but being also imaginary negative. This reflects the double character as being negative (i.e., the negentropy—consuming outsourcing energy) and imaginary—by undermining economical businesses in the environment they act (e.g., by "invisible remote connection" of the negentropic impact;
- *Strategy of animal spirits (triple negative, perfect negative)* presents a value distinct from all other strategic approaches, but it maintains the negative sign of negentropy—as a sign of consumption (generator of economic crises), anticompetitive, antiresources but with immediate action (on short term) for the businesses it characterizes.

Still, ranking issue is still open to these strategic groups, on a unique scale of entropy, based on results (with a meaning and significance), but apparently disjoint at a first comparison, of Table 4.3. To achieve this unifying goal the so-called construction of entropic circles of Figure 4.2 is performed: the real and imaginary (orthogonal) axes joint only in the zero point, while close at "infinite" (away from equilibrium). They would generate two circles of equal radius (regardless the selected radius, they will be equal, consequently with projections identically proportional to the initial axes), which cross after two points generating the so-called "entropic resultant"; on this line, all values in Table 4.3 "successively" arrange themselves, regardless the sign (positive/negative) and the nature (real/imaginary), due to the equal proportions between the axes projected on the circles and the other way round.

The analysis in Figure 4.2 reveals that, actually, the strategy of animal spirits ("bloody competition," specific to "red ocean") is the closest, as strategic entropy to the wise business, due to the specific strategic path, along which, the forces and antiforces, acting during the changes of strategy, annihilate reciprocally; so generating in the end, a minimal entropy in the vicinity of the one specific to the trifecta strategic point (the absolute zero).

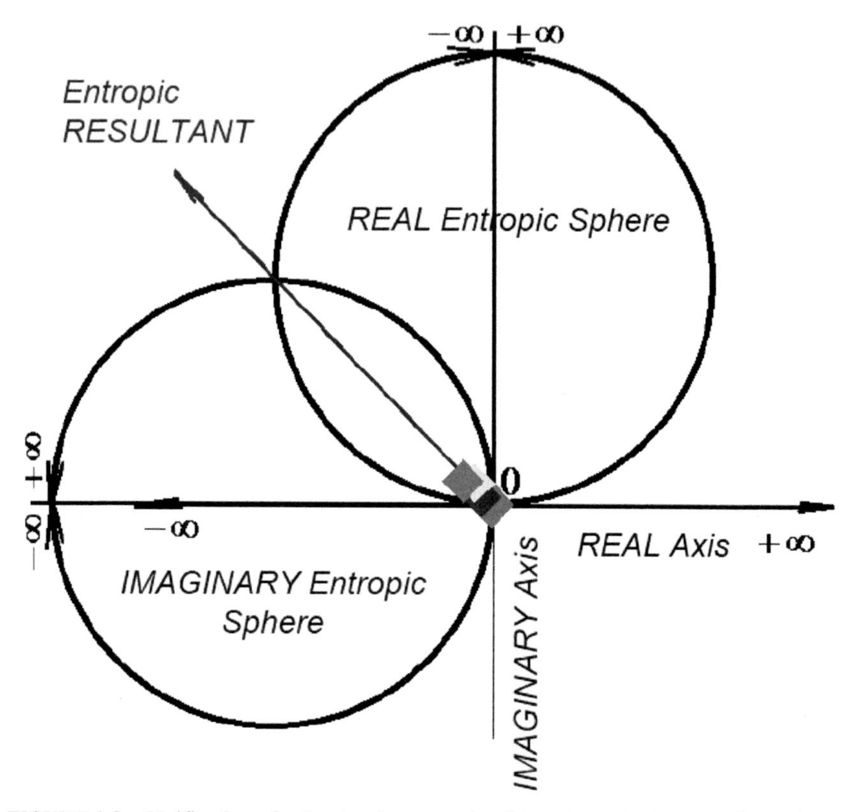

FIGURE 4.2 Unification of entropic spheres, real and imaginary, in the entropic resultant, with a role in ranking results in Table 4.2; the code of the origin's colors (with zero entropy) corresponds to states and strategic groups of Table 4.3, from entropic perspective (S): S (wise business, I) < S (animal spirits, VIII) < S (opportunist-warlike group, II–IV) < S (arrogant group, V–VII).

Thus, it is worth retaining and decoding, of Table 4.2, the strategic path of the spirit type business toward the perfect business of the competitive, sustainable, and regenerative advantage:

Animal Spirits (VIII)

Profit at any economic, ecological and social price → *Polemocracy (VII)*: competitiveness is global, unlike the focus on ecology and social utility → *Smart business (IV):* competitiveness and saving resources are global, but social innovation is ignored →*Outsourcing* (VI): they use low prices for high productivity by saving resources, but with no feed-back/social satisfaction → *Underground business (II)*: profit is not declared as a whole, but there is interest and socio-ecological innovation in developing the business → *Perverse Spirits (V)*: part of the profit stays underground, environment protection is given up, possibly by a campaign to convince the public under the threat of moving away the business, etc. → *Profitable business (III)*: the entire profit is declared, some of it is invested in social programs, but with high consumption of natural and human resources→ *Wise business (I)*: it is achieved, besides integration, in a competitive manner, productive to social utility (on long term), saving ecological means and resources by innovation, so ensuring sustainability of the business development.

Finally, we have to mention that the later differentiation between entropic values of distinctive advantage for strategic manifestations inside the arrogant groups (II–IV) and opportunist groups (V–VII) of Table 4.3 can be done by refining the strategic possibilities in definition (4) so that they would become *conditioned probability*. For instance, each strategy in these groups appears to be next to the animal spirits strategy, as above proved from an entropic point of view; so, by specific path, they all may become the wise business—the trifecta of the competitive, sustainable, and regenerative economical advantage!

4.4 CONCLUSIONS AND PERSPECTIVES

Business states and their strategic approaches are identified in the cube of distinct advantage and considered to be support strategies for the development of strategic paths of innovating a business (product, services).

In a general manner, they are consistently mathematical by combining algebraic information of 3D informational space (through the set of three econo-bits per state/strategic approach) with the longest geometrical paths. This ensures the diminishment of the entropic effect on the business development in an open environment, dynamic and competitive, with limited natural resources, while responding to the high level of public/consumers/society expectancy. The present approach also combines:

- *Complexity management* = through management of minimal entropy (research, interpretation, application)
- *Management of the change* = through the strategy of the optimal path (by coupling min–max in the conjugated characterization, that is, algebraic and informational in closed system and ergodic geometry in an open system).

The results confirm, through specific or group's strategic identifications of the distinct advantage, that "all useful things show similar levels of entropy" Proops (1977), even though apparently the strategy of a group is perceived as contrary. A more accurate identification in subgroups can be performed by taking into consideration the conditioned probabilities on a certain strategic group. This can be done by analogy (constraint or conditioning) to the closest strategy to the perfect one (with zero entropy), herein being identified as the strategy of "animal spirits" (Akerlof and Shiller, 2010): profit by all means, ignoring the crisis of natural resources, cynical attitude toward social challenges and crises.

The present algorithm is also applicable to cases of connection of states in the strategic cube of distinctive advantage when it is linked in the network together with other cubes, neighboring by vertices, edges and faces (representing open flat market, and/or dynamic clusters in expansion, in co-petition, etc.). They are all in relation to the reference (single) strategic cube, marking thus the dynamics of changes in the strategic innovation paths in different contexts as the network strategy. In such cases, the adjacent matrix (vicinity in the network) can be systematically formulated, for which the associated determinant may be calculated. It nevertheless provides information on strategic innovation in estimating the so-called *entropic activities/actions in the network* for a given strategic path. Further on, the maximum of entropic action on a series of connections of strategic innovation paths in networks of strategic cubes will generate the optimal path (maximal) of the innovation strategy for a product/process in an economic network (as several connected markets). Thus, the method has a high degree of generality (Putz, 2019), the present approach representing a preliminary, but fundamental leap, based on informational and entropic considerations, allowing the identification of strategic changes in product/ process innovation, in attaining competitive, sustainable, and regenerative advantage in an original way, along the "innovation paths" away from equilibrium (so minimizing the business entropy).

KEYWORDS

- **strategic cube of distinctive advantage**
- **extended Hamming space**
- **econo-bits**
- **maximum path of the distinctive advantage innovation**
- **entropy hierarchy in the strategic management**

REFERENCES

Akerlof, G. A.; Shiller, R. J. *Animal Spirits: How Human Psychology Drives the Economy and Why It Matters for Global Capitalism*; Princeton University Press: Princeton, 2010.

Georgescu-Roegen, N. Energy Analysis and Economic Valuation. *South. Econ. J.* **1979,** *45*, 1023–1058.

Hamming, R. W. (**1950**) Error Correcting and Error Correcting Codes. *Bell Syst. Tech. J.* **1950,** *29* (2), 147–160.

Haret, S. *Mécanique sociale*, Gauthier-Villars, Paris, 1910.

Ionescu, Gh.; Cazan, E. *Management* (in Romanian); West University Publishing House: Timişoara, 2007.

Marcus, S. *Universal Paradigms* [Originally in Romanian as *Paradigme universale (ediţie integrală)*], Paralela Publishing House: Piteşti, Romania, 2011; pp 169–180.

Nicolescu, B. We, Particle and the World. [Originally in Romanian as *Noi, particula şi lumea*, Editura Junimea, Iaşi], 2007.

Norman, D. A. *On Competition* [*The Psychology of Everyday things*] *The Design of Everyday Things*; Basic Books: New York, 1988; 2002.

Petrişor, I. I. *The Strategic Management. The Potentiological Approach* [Originally in Romanian as: *Management Strategic. Abordare potenţiologică*]; Brumar Publishing House: Timişoara, Romania, 2007.

Pikler, A. G. Utility Theories in Field of Physics and Mathematical Economics. *Br. J. Philos. Sci.* **1954,** *5*, 47–58.

Porter, M. *On Competition*; Harvard Business Review Press: Brighton, 2008.

Proops, J. L. R. Input–Output Analysis and Energy Intensities. *Appl. Math. Model.* **1977,** *1*, 181–186.

Putz, M. V. *Strategic Cube of the Organization Competitive Advantage.* [Originally in Romanian: *Cubul strategic al avantajului competitiv al organizațiilor*]. MBA Thesis, Faculty of Economy Science and Business Administration, West University of Timişoara, 2017.

Putz, M. V. *The Strategic Cube of the Distinctive Advantage. Epistemological Approach*, Chapter 2 of the present monograph, 2019.

Schödinger, E. *What Is Life?* Cambridge University Press: London, 1944.

Simoi, C. *Sociomatics.* (in Romanian); Litera Publishing House: Bucureşti, 1978.

CHAPTER 5

Scientific Entrepreneurship by the Strategic Double Cube of Competitiveness: Knowledge Transfer

ABSTRACT

In the postmodern society, contexts governed by the nanotechnological industry of the multidimensional models of knowledge transfer from the academic/university environment to the entrepreneurship space are analyzed by exploring the hypothesis and actions of the so-called the scientific entrepreneurship. The science (fundamental) is open to the applicative domain by co-participation with extra knowledge to produce the extra value, in the network organizations, especially by clusters or scientific and industrial parks. The analysis of the scientific entrepreneurship with the 2D matrix method reveals four models of action for the associated management (including the nonaction). At their turn, they are retrieved as a particular case for the eight management 3D strategies for the scientific entrepreneurship. The correspondence obtained of the research–development–innovation cube with the inclusive one is expanded on the competitive, sustainable, and regenerative cubic directions while putting in act the postmodern principle of "the corner that becomes the center." The present study and analysis are circumscribed of the sacred versus profane dualism debate for science versus technology. The possible solution seems to be in the competitiveness dual cube (double cube, cube-in-cube), which can finally become a strategic diamond of knowledge exploration (applicative technology) intermediated by the strategic management of 3D information transfer in the new knowledge-based economy/society.

Motto:

"Spray your Truth!
I never Lose!
I either Win or Learn!"

—Campaign RED CODE by STR8 (2016).

5.1 INTRODUCTION

Science or Technology?, Science and Technology?, or Science versus Technology? This is one of the most insidious multilevel questions of the 21st century mainly because of the confusion created around the "revolutions" each component has undergone, respectively:

- Economists speak rather without distinction about "the industrial revolution" of the 19th century. It was eventually turned into "the continuous revolution in techniques and technology" during the 20th century, with a permanent impact on the economic organization (see strategic management), marketing, market (demand and offer), production factors (work, capital and again, management), and so on;
- On the other hand, Kuhn speaks about "scientific revolutions" (Kuhn, 1962; 1970) which induce "cultural revolutions." First of all, since they deal with essential changes of paradigms into "how we perceive the world" (micro and macro, atoms, molecules and constellations, respectively); then, as a *side effect*—also comes the technological revolution. In this last case, techniques borrow from science "as much as it can": the story about Faraday's answer is already famous—the discoverer of electricity to the question "What purpose does it serve? What is the (social) utility?" asked by the board of the Royal British Society on the occasion of being awarded the grant/funds necessary to continue the proposed fundamental researches—to which he answered *"I don't know now, but someday you will certainly charge a fee/tax on this utility!"* And he was so right about it! Or further on, when Max Planck, the discoverer of quantum physics and quanta of light, was sometimes told "Your Excellence, a new application was discovered to your quanta" he used to reply with a sort of indifference "Is that so? That's good!"

In June 2013, during one of the international scientific dissemination tours of the author, as co-organizer and participant to the summer course in the physics and chemistry of carbon in the context of contemporary nanotechnologies (Carbon Topology, 2013), it entered the spirit of the Sicilian town of Erice. This is a town with mythological origins, Erice being the son born from the love between Neptune and Venus; a town with a multimillennial history and worldwide recognized as the town of peace and science. It is noticeable the fact that Erice had then celebrated the 50th anniversary, since the 1961–1963 launching of the Foundation Ettore Majorana and the center for Scientific Culture (2013). This center is well known due to the famous manifesto promoting science as peace mediator as opposed to technology—seen as an instrument of the politics of power and, consequently, of destruction (virtually or not) of humanity. The original title of this manifesto was "*Science is the study of Fundamental Laws of Nature. Technology is the study of how the power of mankind can be increased,*" signed in 1982 by the famous Nobel Prize for Physics laureates Paul A. M. Dirac and Peter Kapitza, together with the Nobel nominee Antonino Zechichi—astrophysicist, excellent organizer, and debatable physicist involved in the history of great subnuclear discovery of CERN. The Majorana center, with a prodigious activity of over 50 years of activity, over 124 schools, 1573 courses, 112,278 participants, and 128 Nobel Laureates of not less than 932 universities and laboratories from 140 countries and nations, brought his contribution to a true international emulation, the so-called Erice spirit (*Erice Geist*). It includes even the setting of an international Foundation of Scientists—The World Federation of Scientists (WFS 2013), which is currently fighting for the 15 planetary emergencies (i.e., *water, soil, food, energy, pollution, limiting development, climate changes, global monitoring of the planet, military threat in a multipolar world, avoiding north–south ecologic holocaust for developing countries through science and technology, the issue of organ substitution, infectious diseases, cultural pollution, common defense against cosmic objects, huge military investments*). Even the Pope John Paul II (1993) visited the center and its activities and made the decisive sentence by admitting that "Science and Belief are both gifts of God" (the original version—"*Scienza e Fede sono entrambe doni di Dio*"), as stated in the third Phrase to WFS during the visit to Erice in 1993. In the same spirit is also circumscribed the famous VIII Phrase "Man may perish through the effect of technology he himself had developed, and not because of the truths he discovers through scientific research," in original "*L'uomo può perire per effecto della tecnica che egli stesso sviluppa, non della verità che egli scopre mediante la ricerca scientifica.*"

But nevertheless, there is a conciliation solution of *Science*—so close to the fundamental problems of man (namely, Where does he come from? What is he made of? Where does he go?—*Qvo Vadis Hominis*?) and the *Technology*—as an intermediate between body and soul/spirit, between wish/goal and achievement/creation of added value, for himself and for the society, as well? Potentially by practicing the Aristotelian method—*Eudaemonism*—brings happiness (including prosperity of the individual and social insertion)? There is a way! It is a truth that does not come from *economic planning* (already dated in the age of networking and leveling traditional organizational ranking)—but rather from the *strategic management* (already autonomously developed as a paradigm of procedural rationality, in contrast with the substantial one, oriented to objective-profit, economy-efficient type). It is about of the so-called *transfer of knowledge* between the academic and research world to economic agents, organizations, and corporations owning market data and developing markets. It is about creating new needs appropriate to the added value patented from their own research/investment, acquisition, and development of patents—the *know-how*. It is about attracting human capital and intelligent data by the knowledge transfer, discontinuous and disruptive, while yet not systematic, respecting the human and ecological global needs.

Knowledge transfer comes from the scientific side (subtly different from technological *know-how*, connected to ownership of patents, licenses, franchises, trainings, etc.), to the current production sector—*renamed* (from industrial perspective) *as nanotechnological* (perhaps due to the customized production feature, specific, on-demand, and so on, and often complementary, though many times and in many cases priority to mass production). On the other hand, *the activities associated to this knowledge transfer*—are specific to conjunct systems in order to create and develop value, thus oriented *to generalizing the chain of value to the constellation of value, sometimes called*—(though somehow limitation) scientific entrepreneurship (or *research entrepreneurship*). They both can represent the viable solution (sustainable, thus) *of the reconciliation of science and technology in the framework of new patterns of sustainable business* (ensuring the competitive, sustainable, and regenerative advantage in the context of *new economy* as *network economy/share economy*).

Such a goal is approached in the present essay-study being conceptually inspired by the assumptions of the recent investigation (with quasi-statistic validation due to the particular analysis at the mono-organization level) linked to the transfer of scientific knowledge in nanotechnology (Zalewska-Kurek et al., 2016), as presented in the Cognitive Analysis section. The author's methodology development is unfolded in the Methodological Analysis section of classifying/approaching strategic management in the cube of competitive, sustainable, and regenerative advantage (Putz, 2017, 2019). Accordingly, a *mutatis-mutandis* design with the *cube of research–development–innovation*, and, respectively, of the management of scientific entrepreneurship is advanced.

The result of this triple juxtaposition generates particular patterns of sustainable business (with scientific control of the manifested technology), which practically widen, but include as well the work hypotheses of the group of authors (Zalewska-Kurek et al., 2016).

5.2 COGNITIVE ANALYSIS

Zalewska-Kurek et al. (2016) identified four matrix modes and three functional modes for scientific knowledge transfer into nanotechnology; then they tested them through three hypotheses and concluded, by a strategic positioning the "scientific differentiation" for an optimal result (unfortunately nonconfirmed by the study in question) of scientific entrepreneurship. However, their approach reveals certain conceptual, cognitive, interesting, and catalyzing passages for a general theory of the strategic development of the research–development–innovation potential in nanotechnological (meta)-clusters (Putz, 2016–2019), which will be illustrated in Table 5.1 "face-to-face."

5.3 METHODOLOGICAL ANALYSIS

Postmodern management is necessarily multidimensional, at least 3D, so that "it would get out" of the fetish matrix of planning industrial management (in relation to mass production).

TABLE 5.1 Conceptual Development of Context, Hypotheses, Modes, and Activities of Transfer of Scientific Knowledge in Nanotechnology According to Zalewska-Kurek et al. (2016) in the Critical Constructive Presentation of the Author's Concept (Putz, 2016–2019).

Ideas, concepts, and developments (Zalewska-Kurek et al., 2016)	Observations, notes, and critical analysis (Putz, 2016–2019)
General aspects	
• *Third mission of university*	• *Entrepreneurship, after education, and research*
• *Infosphere of knowledge domain: patents, (new venture creation) and support—including innovation network creation and so on—see Rothaermel et al. (2007)*	• Attributes of nanotechnological clustering
• *Management of networking, including patent, spin-offs, research contracts, training courses in industry, consultancy, research grants, oriented university graduation, knowledge parks, and so on—see Bekkers and Bodas Freitas (2008)*	• Nanotechnological clustering activities
• *Knowledge transfer, co-engagement with industry, co-engagement collaboration, active co-creation), and so on (see Perkmann et al., 2013)*	• Attributes of scientific entrepreneurship
• *What are the conditions that make scientists explore and exploit the opportunities of the alliance with industry? What are the management conditions to optimize knowledge transfer in collaboration with industry* [in the broad sense of nonacademic organizations]?	• Key questions to create and develop scientific entrepreneurship
Patterns, hypotheses, and strategic positioning	
• *Relations between the scientific world and the industrial world (contemporary nanotechnological) has in general a temporally determined nature; because scientists are mainly interested in knowledge production, its dissemination in view of career development, actions not necessarily amplified by a relationship to industry. (van Rinnsoever et al., 2008)*	• Challenge of clustering, in the sense of providing a competitive, sustainable, and regenerative advantage to the "university world" as well, engaged in scientific entrepreneurship!
• *Possibility to develop a temporary strategic alliance, potentially in an organizational integration* process [not necessarily absorption, case in which a research institute, for example, would become a research development in a company, n.a.]	• Clustering strategy, with temporal dynamics (*in and out*, based on achieving objectives and develop common projects)

TABLE 5.1 *(Continued)*

Ideas, concepts, and developments (Zalewska-Kurek et al., 2016)	Observations, notes, and critical analysis (Putz, 2016–2019)
• *Strategic alliances develop competitive co-advantages (Perkmann and Walsh, 2007)*	• Specific to maturated clusters, in terms of a sustainable advantage at the network level
• *The vulnerable strategic positions are characterized by the need of resources; the strategically strong positions are characterized by the need of sharing valuable resources; consequently, every partner, in any combination, is in the possession of heterogeneous and exchangeable value/resource, in whose absence the strategic alliance is not possible (see Gueth, 2001)*	• It affirmed the principle (even the theory) of economic asymmetry in relations, and in clustering at a more general level
• *[Inter] organizational integration based on strategic interdependency, by pooling unevenly distributed competences and resources*	• Attention should be paid to the culture, purpose, and strategy of the organizations involved! They have to be made compatible in network type relationship (with intense relationships and knowledge transfer
• *The two factors responsible with producing value in co-organization/strategic alliance are as follows*: (1) need for strategic interdependency sharing [asymmetrical] resources and (2) need for autonomy (*see* Haspeslagh and Jemison, 1991)	• Possibly scientific knowledge is shared upstream and downstream the "product" of the industrial technology, and non-patent dissemination of the university is differentiated
• *Research management with knowledge transfer activities seen as dependent variables (see Zalewska-Kurek et al., 2010), Lyhne, 2012)*	• Management practices/techniques can be regarded as function of independent variables, asymmetrical resources and intra-organizational autonomy: $\frac{KNOWLEDGE}{TRANSFER} = f_{\substack{RESEARCH \\ MANAGEMENT}}(RESOURCES, AUTONOMY)$
• *Management modes correlated to organizations' behavior* 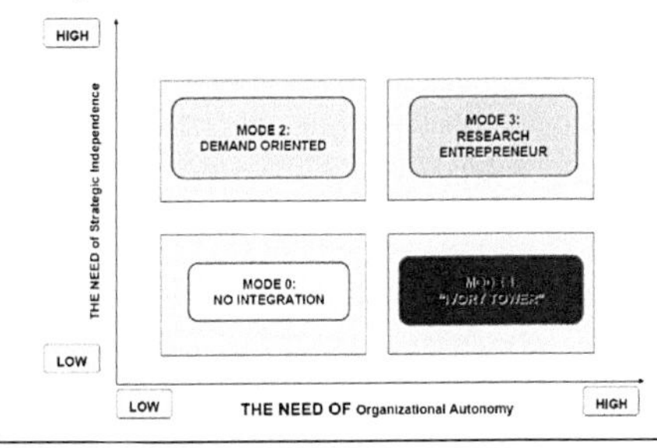	• One specializes the function f^{RESEARCH} MANAGEMENT by a matrix; however, in the spirit of modern management (namely, planning) and not in the spirit of post-modern management (namely, knowledge spiraling, i.e., of organizing the chaos", actually by the multitude of data and of the eight-fold informational superposition of fuzzy- intra and extra organization); in a certain sense, the functional

TABLE 5.1 *(Continued)*

Ideas, concepts, and developments (Zalewska-Kurek et al., 2016)	Observations, notes, and critical analysis (Putz, 2016–2019)
	dependence can be conceptualized as $KNOWLEDGE = f_{\substack{RESEARCH \\ MANAGEMENTI}}(ENTREPRENEURSHIP)$ More generally (and in a much bolder way), it may taking the philosophical level form as $REGENERATION = f_{\substack{SCIENTIFIC \\ MAANAGEMENT}}(LIFE)$ with all ethical aspects involved (e.g., genetically modified organisms, cloning, genetic engineering, etc.).
• *Zero mode is not considered to be interesting for the discussion, so it is neglected by* Zalewska-Kurek et al. (2016)	• Still, *zero mode* can be associated to the so-called "zero relationship" (in the sense of nonnetwork) of the relational marketing promoted by Gummerson (2006); Furthermore, the zero mode will correspond to the technique of *interface management* present in the author's approach, presented in the next section in the context of the strategic cube of competitiveness
• *Mode 1 is associated to fundamental research, purely academic ("ivory tower") and without connectivity motivation, followed by knowledge transfer (also see Gibbons et al., 1994)*	• In the strategic deployment of a business, the pattern/model of "ivory tower" corresponds "to animal spirit" type entrepreneurship. Actually, the selfish accumulation thorough *knowledge management* enhances the informational asymmetry and the competitive advantage, see the author's approach, in the context of the strategic cube of competitiveness (see the next section)
• *Mode 2 ("orientation to demand/market") is of corporatist type, focused on conquering markets, possibly with political and governmental support—especially in relation to direct foreign investments of certain multinational corporations (see as well Ziman, 1991)*	• The model oriented toward on market demand corresponds to businesses based on strategic marketing, or marketing intelligence, or more generally known as *marketing management*. It is a persistent exploitation of the existing resources (possibly by monopolization of new ones), namely, the model of polemo-centric business; see more considerations in the next section presented in the context of the strategic cube of competitiveness

TABLE 5.1 *(Continued)*

Ideas, concepts, and developments (Zalewska-Kurek et al., 2016)	Observations, notes, and critical analysis (Putz, 2016–2019)
• *Mode 3 ("scientific entrepreneurship") would co-opt asymmetric resources in a network type collaboration, with compliance for the need for strategic interdependence and for intraorganizational autonomy for the collaborating entities*	• The model of "scientific entrepreneurship" correlates with competition (cooperation + competition) between participating organizations (even temporarily) in a meta-organization (cluster type). It features strong strategic characteristic, thus belonging to the strategic management (*win–win* type), so allowing the technique of *predesign management as a complete strategic deployment*); see it presented in the context of the strategic cube of competitiveness in the next section
• *H1 validation hypothesis for knowledge transfer: the greater the need for organizational autonomy and strategic independence, the bigger the likelihood of knowledge transfer*	• Corresponds to *model 3* of "project management" within the scientific entrepreneurship, consequently the hypothesis is apparently redundant, yet assertive! In symbolic terms of statistical correlation factors, we have: $R_{f(V1,V2)}\uparrow \vert \{R_{V1}\uparrow \& R_{V2}\uparrow\}$
• *H2 validation hypothesis for knowledge transfer: orientation to an industrial career [of corporate type] has a positive effect on the need for strategic independence and organizational autonomy through the knowledge transfer activities performed*	• This hypothesis corresponds to the model 2 of "marketing intelligence/ management," yet hiding a small contradiction in the authors' approach (Zalewska-Kurek et al., 2016). Actually, it does not conceptually correlate with the matrix definition of model 2, by which it should be characterized by lower autonomy to the affirmation of its growth (as a positive impact through knowledge transfer) in the test hypothesis; thus, one may anticipate its invalidation (also concluded by the authors, as well) upon a testing analysis
• *H3 validation hypothesis for knowledge transfer: orientation to an academic career negatively affects both the need for strategic interdependence and the need for organizational autonomy through the knowledge transfer activities performed*	• Zalewska-Kurek et al. (2016) pre-correlate the high need for intraorganizational autonomy [perfectly valid when it comes to universities] from the definition of model 1

TABLE 5.1 *(Continued)*

Ideas, concepts, and developments (Zalewska-Kurek et al., 2016)	Observations, notes, and critical analysis (Putz, 2016–2019)
	• (ivory tower), to which corresponds hypothesis 3, with the assertion of negative impact on this autonomy through the knowledge transfer activities; however, being a negative hypothesis, it is harder to validate in anticipation, because of false positives (the lack of observation does not mean lack of the objective to be observed)!
Measurement—quantifying indicators	
• *The main identified activities of knowledge transfer are*: (1) *no activity (the model 0)*, (2) *exploration of a collaboration intention (the model 1 which only recognizes virtual collaboration opportunities)*, (3) *preparation of collaboration through negotiation (the model 2, by market strategic intelligence), and* (4) *exploiting collaboration by contracts with the nano-technological industry (the model 3, by spin-offs)*	• Much more main activities for each model, as well as the expansion of knowledge transfer models will be presented in the next section, in the author's vision, with the method of the competitive strategic cube
• *Indicators of the need for strategic independence*: *access to knowledge and competences, access to research facilities, enhancement of scientists' reputation, of the scientific expertise, networking, commitment, employability (percentage of the university graduates with a job in industry) and social capital*	• There is missing the connection to the ecologic environment, need for sanitary, social safety, securing future economic sustainability, global peace, and sustainable progress
• *The indicators of the need for organizational autonomy—objective type are*: *assuming the role of the project leader, the project decision-making capacity (including knowledge about the contributions from industrial partners), the academic position (level), the academic status (the determined/ undetermined period of time and the title obtained—PhD, postdoc, entitled, etc.), the seniority [degree of excellence and possibly function]*	• It could be added self-determination in choosing the working hours, of the time for projects, intra-academic relationships, project bonuses and salary min–max grid according to the assumed/performed performance, and so on

TABLE 5.1 *(Continued)*

Ideas, concepts, and developments (Zalewska-Kurek et al., 2016)	Observations, notes, and critical analysis (Putz, 2016–2019)
• *Indicators of the need for organizational autonomy—subjective type: freedom in the decision on reporting/research products (deliverables) [non] restriction in the choice of research and publications, meeting [multi-role] or all-roles (employee–employer–manager) in a partnership project university-industry*	• We may add the propensity for individual cooperation agreements individual academic researcher-industry, the role of scientific advisor for industry, subsidies to equip research laboratories and/or didactical academics from collaborative projects university industry, employability of their own doctoral/postdoc in the industry they have developed partnerships with knowledge transfer, and so on
• *Indicators of the industrial orientation [corporatist]: the objective to develop spin-off, previous or current experience [in collaboration] industry, intention to work [part-time or full-time in the near future] in industry*	• You might need to add third-party contracts, wage growth, econo-ecological involvement, need for career diversification by enforcing fundamental research, and so on
• *Indicators for academic orientation: the desire to develop a research group, desire for excellence in education, interest in advanced research in the field, the motivation to interact with students, appetite for scientific challenges—new directions [and emerging areas], motivated by the academic freedom*	• The need for the independence within the university, self-sufficiency, exclusive inclination toward conceptualizing, theorizing, reflective research, even "civic fatigue" or refusal of serving technology, religious beliefs, and so on
• *Interviews conducted to investigate the observed reality (by open-ended interviews) with quantifying: 0—for the "not-Observed," "+1"—for the "low Observed Needs," and "−1"—for the "high Observed Needs"; the statistical analysis of the main components type with the need for interdependencies was undertaken and the 1D and 2D results needed a range of validity [as anticipated, e.g., the subclassification according to the above objective and subjective indicators]*	• The quantification of the values in the 3D Hamming–Putz space of values {−1,0,+1} can catalyze further studies related to the action of strategic forces of academic versus industry competition (Putz, 2016–2019); but it also fundaments the extension of the statistical analysis, in best case for the two-dimensional—namely, the principal component analysis results in 3D space of the strategic cube of competitiveness, see the next section
Reproblematization	
• *Including in the statistic analysis the academic rank as a distinct explanatory variable*	• It forces analysis by preselection of an indicator for the need for autonomy responsible for the entire set of autonomy indicators (be it only objective)

TABLE 5.1 *(Continued)*

Ideas, concepts, and developments (Zalewska-Kurek et al., 2016)	Observations, notes, and critical analysis (Putz, 2016–2019)
• *It is concluded that academic rank explains model 3 of scientific entrepreneurship*	• We believe this is an error/confusion of Zalewska-Kurek et al. (2016) academic rank better (and directly, by default) correlating with Model 1 of the "ivory tower"—full academic autonomy, according to their own behavior modeling scheme for knowledge transfer—see above
• *Zalewska-Kurek et al. (2016) reject hypotheses H2 and H3 (above) as not being validated by their own study*	• If we have agreed in advance that hypothesis H2 will be invalidated due to "its weak formulation, being in its own hidden contradiction with the model assumed—see above," we believe that hypothesis H3 is equally wrongly rejected. This conclusion comes exactly from its being formulated in "the negative," inducing false positives, and again directly anticipated: the academic orientation is indeed a negative impact on the course of knowledge transfer to industry. In addition, the authors are "trapped in this confusion" also because of their previous confusion which correlates academic rank to scientific entrepreneurship (confusion between models 1 and 3)
• *Models of Zalewska-Kurek et al. (2016) are designed as based on the classification of Haspeslagh and Jemison (1991), namely,* (1) *The Model 1 is of the "ivory tower" and associates with type of preservation of the strategic position* and (2) *The Model 2 of the "marketing management" is associated with absorption strategy [of resources]; while the Model 3 is of "scientific entrepreneurship" and corresponds to a symbiosis strategy*	• These models can be integrated into a higher stage (generalized) of analysis in the so-called chrysalis economy, namely, Crysalis Economy (Elkington 2001), noting the fact that of all the natural sciences Biology is closest to the dynamics of economic processes (namely, life cycle–product, clustering–swarm, competition–survival, adaptation–alliance, evolution, social groups, hierarchies, and organizational chaos)

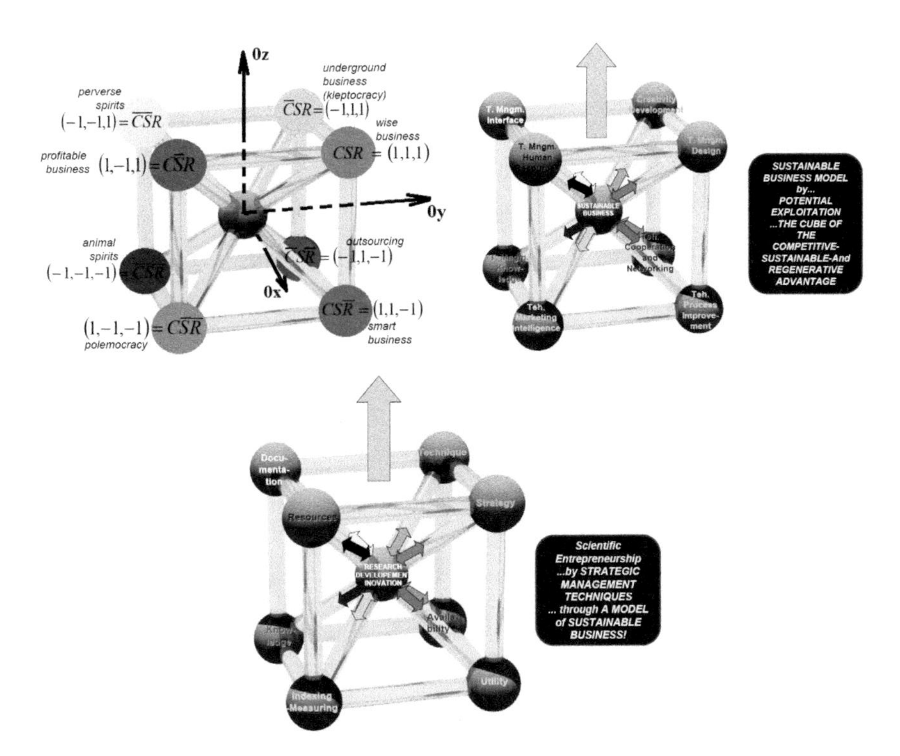

FIGURE 5.1 The Sequence of strategic cubes, from top to bottom: the cube of competitive, sustainable, and regenerative advantage in 3D deployment. It covers as follows: the positive–negative space of business patterns in compact network (on top); the cube of adjacent management techniques (creator of) sustainable business, in the sense of supporting "in equal potentials—nonranking" of it, respectively to the competitive, sustainable, and regenerative advantage (in the middle); and the cube research–development–innovation in 3D compact expansion (so covering the space of knowledge—with no gaps) with all its components identified correspondingly to the sustainable business cube (on bottom). The three cubes are in circular relationship (from bottom to top as in the connotative correspondence, and from top-to-bottom as necessity and denotative correspondence) in the context of knowledge transfer of scientific entrepreneurship.

In the cognitive analysis of the recent approach of knowledge transfer from the scientific community (academic, university, and research institutes) to the technical-applicative one (generically individualized as nanotechnology in the postmodern era)—a slightly imperceptible mix between modern and postmodern has been registered in the approach of Zalewska-Kurek et al. (2016).

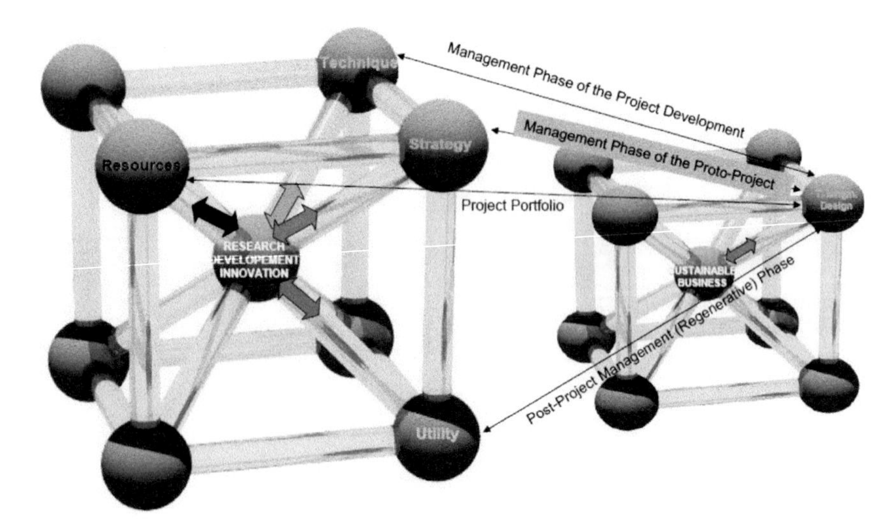

FIGURE 5.2 "Interstitial" correspondence for the techniques of *innovative project management*_in the scientific entrepreneurship through knowledge transfer between spaces (cubes) and the selected potentials of research–development–innovation and sustainable business.

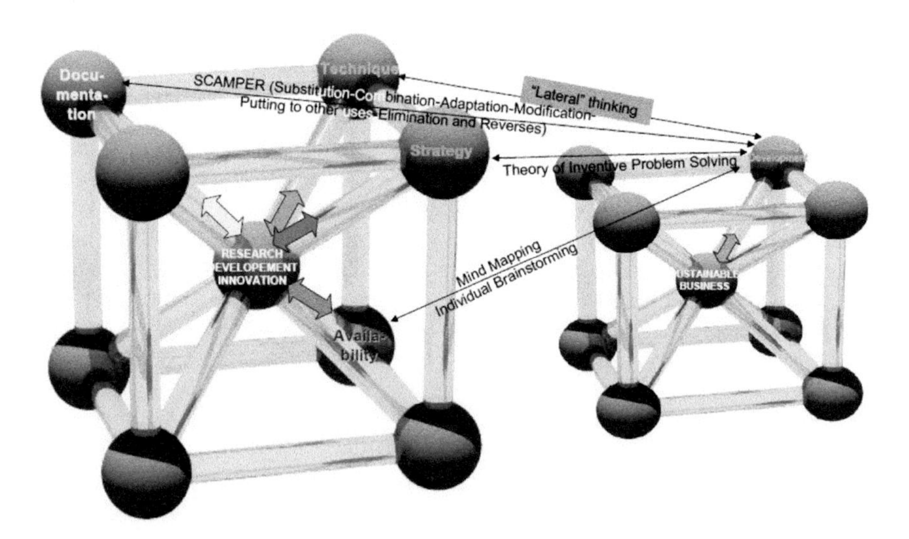

FIGURE 5.3 As in Figure 5.2, here with "interstitial" correspondence for techniques of the creativity development management.

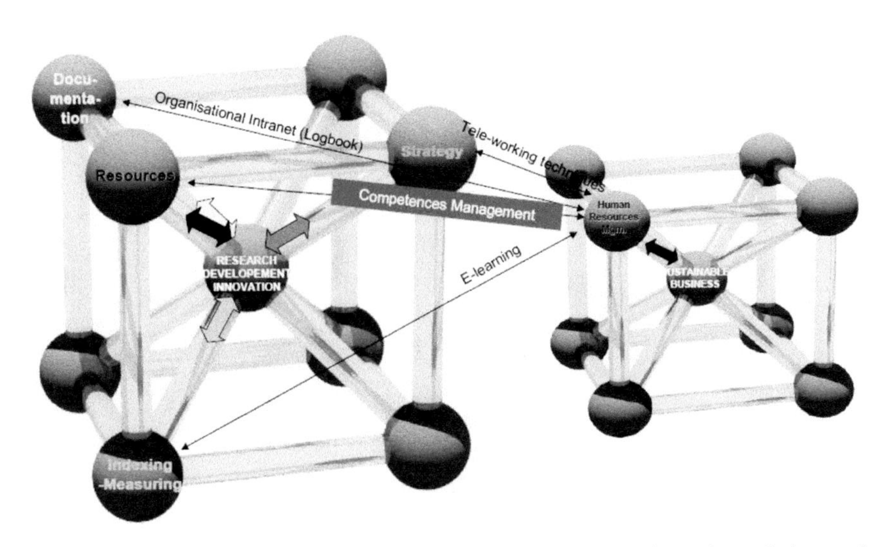

FIGURE 5.4 As in Figure 5.2, here with "interstitial" correspondence for techniques of *human resources management.*

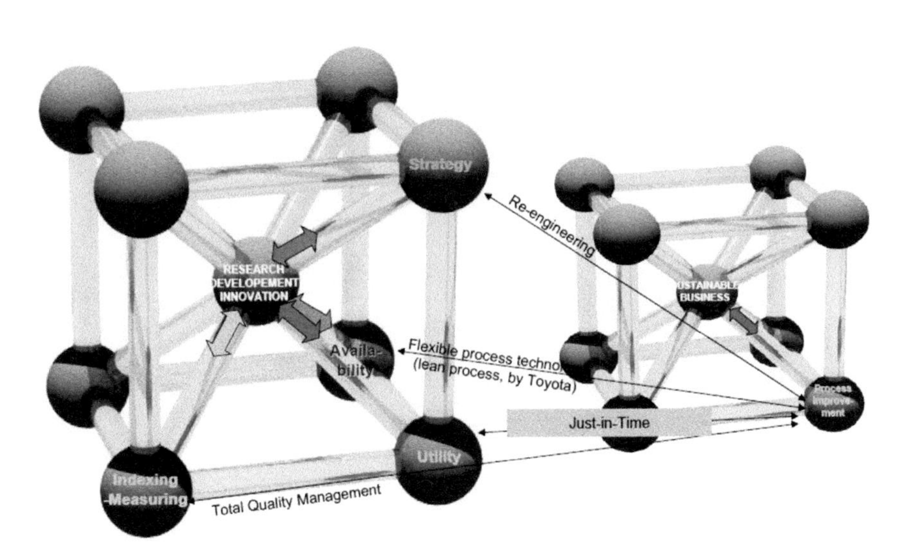

FIGURE 5.5 As in Figure 5.2, here with "interstitial" correspondence for techniques *of total quality management (process improvement).*

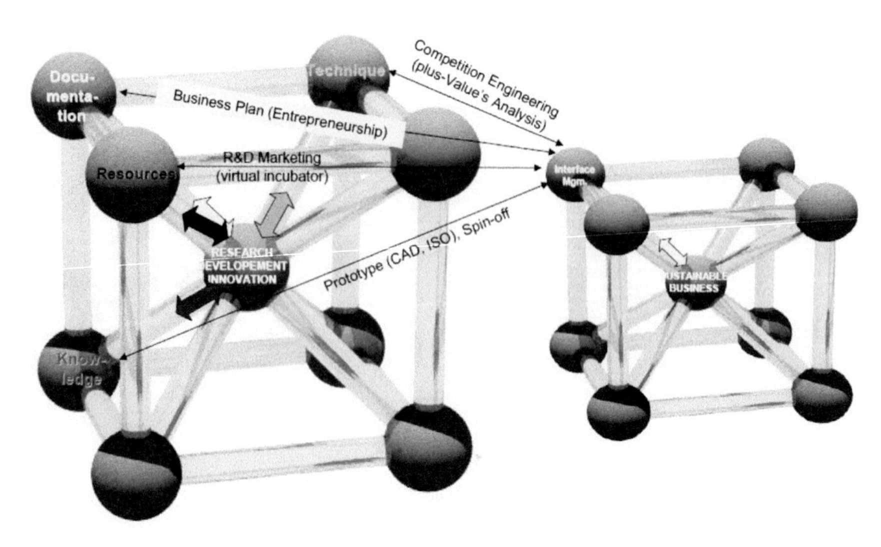

FIGURE 5.6 As in Figure 5.2, here with "interstitial" correspondence for techniques of *interface management (infosphere).*

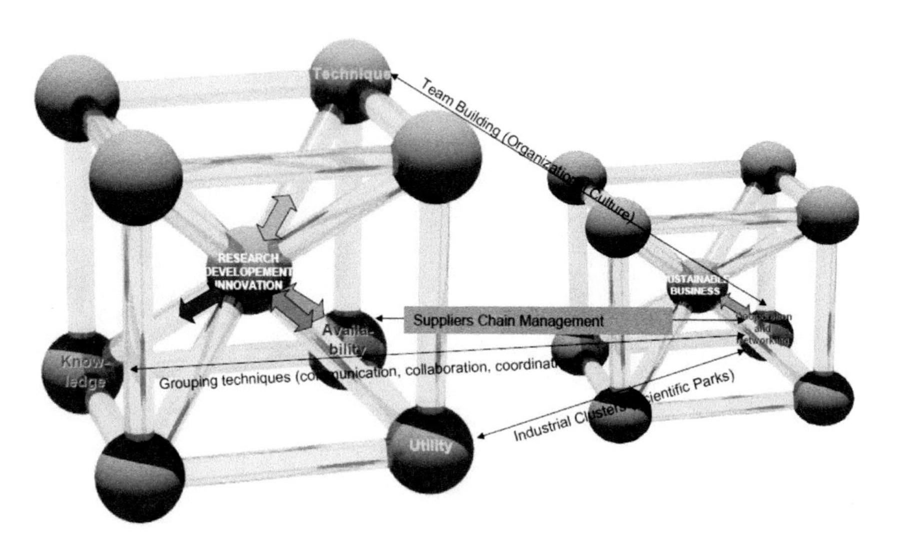

FIGURE 5.7 As in Figure 5.2, here with "interstitial" correspondence for techniques of *cooperation and networking management (coopetition and networking).*

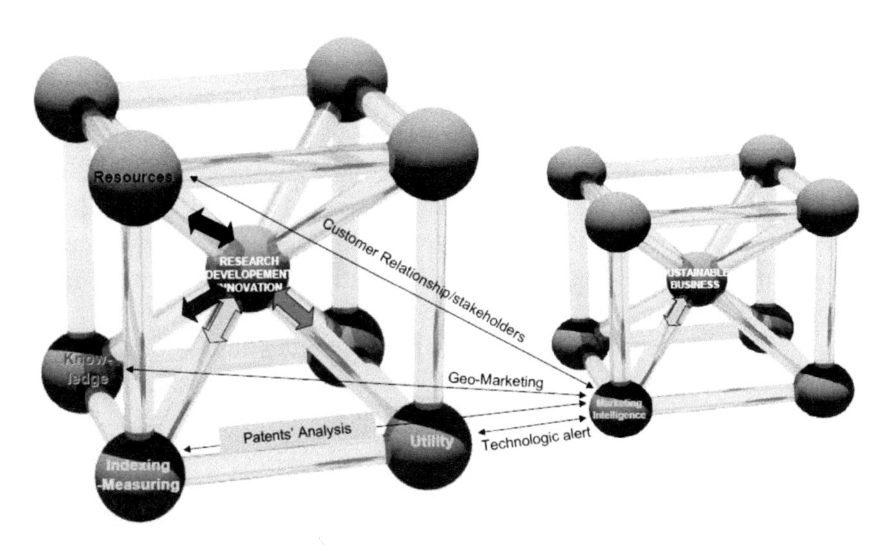

FIGURE 5.8 As in Figure 5.2, here with "interstitial" correspondence for techniques of *marketing and intelligence of the market management.*

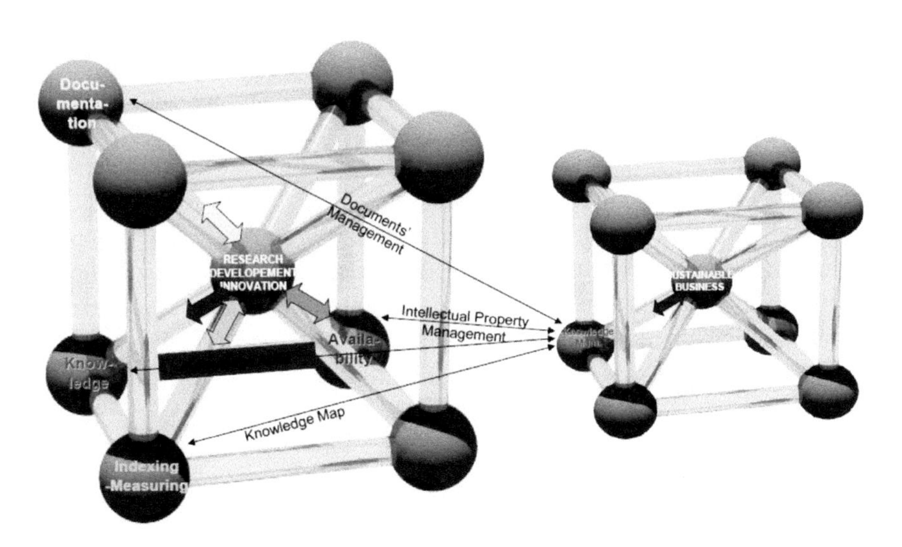

FIGURE 5.9 As in Figure 5.2, here with "interstitial" correspondence for techniques of *knowledge management.*

Hence, the fact that now and then the partial *postmodern* (3D) management conclusions are not very consistent with the presumptions and hypotheses of the *modern* (2D) management strategic analysis. For example, four strategic positioning methods have been introduced through method "2D-high/low" in modern management, but there were actually used three independent variables (the need for strategic independence, objective autonomy and the need for subjective autonomy)—as it resulted from the analysis of the main components in the statistic study performed; not to mention that this one was also carried out for just one single organization, eminently experienced in knowledge transfer and with a rich portfolio of nanotechnology, that is, at "MESA+"/Institute for Nanotechnology at the University of Twente, the Netherlands. Thus, the question of self-consistent approach, consistent postmodern, consistent 3D arises, by potentially recovering but also enriching the strategic patterns already presented in the activities of knowledge transfer. Fortunately, this conceptual context is available to the present author; it refers to the *strategic cube of competitiveness*, with the variants (Fig. 5.1) as follows:

i) The cube of competitive, sustainable, and regenerative advantage, with forms of businesses, adjacently resulted.
ii) The cube of sustainable business, with managerial adjacent forms.
iii) The cube of research–development–innovation, with entrepreneurship adjacent manifestations.

All these cubes have corresponding ideas "corner-to-corner," with possibility of conceptual development resulting from reciprocal "mapping." They thus are providing, in a geometrical–conceptual manner, the specification of the *transfer function* for the management of the scientific entrepreneurship. This function should resemble the function which "made the passage" from entrepreneurship to knowledge in the cognitive analysis (Table 5.1):

$$Knowledge = f_{research\ management}(Entrepreneurship),$$

but which actually ensures the development of a certain type of business able to provide added knowledge by scientifically managing the entrepreneurship added value as input.

In order to understand and customize the types of scientific entrepreneurship allowed by the method of strategic cubes (of sustainable business with competitive, sustainable, and regenerative advantage compared to that of research–development–innovation) we have to mention the construction "philosophy" of the original strategic cube, see (Putz, 2017, 2019/Fig. 2.3):

i) It is considered the Hamming–Putz space, mentioned above, {−1,0,+1}, consisting of strategic positioning of non-observable type "0", observable with positive impact "+1" and with negative impact "−1." Maintaining the non-observable characteristic in the center of the space (informational, economic, business, research, etc.), the possible "3D steps," as combinations of positive/negative characters in triplets (a, b, and c), lead to eight-strategic points laid out around the nonobservable strategic point. They are "expanded" or manifested toward it in the corners of a strategic cube. They are the strategic equivalent of "Big-Bang" in astrophysics or cosmology: the nonobservable "gets out of itself" and fills in the space of existential possibilities (generating strategic points). In the Hamming–Putz space of information (stepping information) the generating cubes can be self-contained in each other (see also the Conclusions section), but also as adjacent cubes—with common corners, edges, and sides. We will deal with the latter case on a different occasion (regarding the strategic approach of *logistics management*, see next Chapter). Nevertheless, here we paid special attention to the notion of *the cube-in-the cube*, in relation to the "mapping" spirit we had mentioned before, that is between the cube of competitive, sustainable, and regenerative advantage and that of research–development–innovation.
ii) According to every strategic point we get, depending on the number of positive characters vs. negative ones present in its name/codification, the corresponding type of business is identified: from the business mainly positive (+1,+1,+1) = wise business, to the business mainly negative (−1, −1, −1) = animal spirits.

By the circular connotative-denotative method ("cube to/in cube") we identify the "strategic points/strategic positions" both in the space of research–development–innovation and sustainable business management (durable at the level of the organization while sustainable at ecological, social and environmental level), see Figure 5.1—middle and bottom. By using this methodology, we are able to identify, based on the same principle of "mapping," the 3D coordinates/behavior along the three independent directions Hamming–Putz, around every strategic point/position, corresponding to a type of business/activity/scientific entrepreneurship management.

iii) A connection is settled between two equivalent positions of the two strategic cubes, the research–development–innovation, and the sustainable business…"mapping" stage.

iv) The cube of research–development–innovation is considered as "the big cube," as it is observed by the scientific entrepreneurship behavior. The propensity for an academic, university, or research organization would "open" to the economic area. This transferring principle is essentially targeted by identifying the knowledge transfer mechanism from fundamental/scientific knowledge to applicative/technological/industrial nanotechnology, including its feedback in the opposite position as well.

v) From "the big" cube of research–development–innovation the Hamming–Putz type "informational steps" are identified, by moving onto each of the three orthogonal directions, thus independent, centered in the corner/strategic position of interest. The new strategic positions will be taken as independent variables in the central function. In this "process," the "corner becomes center" in line with the inclusive characteristic of postmodern approach according to which "periphery becomes the center!" The result is the scientific entrepreneurship of the same type as the identified business from the corresponding central corner of the cube of the competitive, sustainable, and regenerative advantage. The strategic management is of the same type with the management identified in the corresponding corner of the sustainable business cube.

In a synthetic manner, we have for each case, the sheet of the corresponding strategic mode:

$$BUSINESS\ \ MODEL \supset$$

$$TYPE\,/\,SCIENTIFIC\ \ ENTREPRENEURSHIP\ MGM. = f_{\substack{RESEARCH\\DEVELOPMENT\\INNOVATION\\CUBE}}\begin{pmatrix} V1_{COMPETITIV}(a1), \\ V2_{SUSTAINABLE}(a2), \\ V3_{REGENERATIVE}(a3) \end{pmatrix}$$

Specifically, Figures 5.2–5.9 put into practice the present methodology, with functional results (as function of functions) subscribed in a cassette of the respective strategic mode/position, fully characterized:

TYPE OF BUSINESS =...

MODE/SCIENTIFIC ENTREPRENEURSHIP MGM =...

Scientific management function =...

Variable **V1** (of competitiveness) =...

Activity *a1* (in service of competitiveness) =...

Variable **V2** (of sustainability) =...

Activity *a2* (in service of sustainability) =...

Variable **V3** (of regeneration) =...

Activity *a3* (in service of regeneration) =...

For Figure 5.2, we have MODULE-1 of scientific entrepreneurship with knowledge transfer:

TYPE OF BUSINESS = *smart business*

MODE/SCIENTIFIC ENTREPRENEURSHIP MGM = *predesign*

Scientific management function = *Strategy*

Variable **V1** (of competitiveness) = Technique

Activity *a1* (in service of competitiveness) = *Design Development Management Stage*

Variable **V2** (of sustainability) = *Resources*

Activity *a2* (in service of sustainability) = *Project Portfolio*

Variable **V3** (of regeneration) = Utility

Activity *a3* (in service of regeneration) = *Post-Project Management Stage* (regenerative)

For Figure 5.3, we have MODULE-2 of scientific entrepreneurship with knowledge transfer:

TYPE OF BUSINESS = *underground business (kleptocracy)*

MODE/SCIENTIFIC ENTREPRENEURSHIP MGM = *"lateral" thinking*

Scientific management function = *Technique (existent and necessary know-how)*

Variable **V1** (of competitiveness) = *Strategy*

Activity *a1* (in service of competitiveness) = *Inventive Solution to problems*

Variable **V2** (of sustainability) = *Documentation*

Activity *a2* (in service of sustainability) = *SCAMPER (Substitution–Combination–Adaptation–Modification–Putting to other uses—Elimination and Reverses)*

Variable **V3** (of regeneration) = *Availability*

Activity *a3* (in service of regeneration) = *Mental Design (Mind Mapping)/ Brainstorming Individual)*

For Figure 5.4, we have MODULE-3 of scientific entrepreneurship with knowledge transfer:

TYPE OF BUSINESS = *profitable business*

MODE/SCIENTIFIC ENTREPRENEURSHIP MGM = *Competences Management*

Scientific management function = *Human Resources*

Variable **V1** (of competitiveness) = *Documentation*

Activity *a1* (in service of competitiveness) = *Organizational intranet (Log)*

Variable **V2** (of sustainability) = *Strategy*

Activity *a2* (in service of sustainability) = *Tele-work techniques*

Variable **V3** (of regeneration) = *Measurement*

Activity *a3* (in service of regeneration) = *E-learning*

For Figure 5.5, we have MODULE-4 of scientific entrepreneurship with knowledge transfer:

TYPE OF BUSINESS = *smart business*

MODE/SCIENTIFIC ENTREPRENEURSHIP MGM = *Continuous and on time process improvement*

Scientific management function = *Utility*

Variable **V1** (of competitiveness) = *Availability*

Activity *a1* (in service of competitiveness) = *Technology of the flexible process (lean process by Toyota)*

Variable **V2** (of sustainability) = *Measurement*

Activity *a2* (in service of sustainability) = *Total Quality Management*

Variable **V3** (of regeneration) = *Strategy*

Activity *a3* (in service of regeneration) = *Re-engineering*

For Figure 5.6, we have MODULE-5 of scientific entrepreneurship with knowledge transfer:

TYPE OF BUSINESS = *perverse spirits*

MODE/SCIENTIFIC ENTREPRENEURSHIP MGM = *Interface management*

Scientific management function = *Documentation*

Variable **V1** (of competitiveness) = *Resources (available and necessary)*

Activity *a1* (in service of competitiveness) = C&D Marketing (virtual incubator)

Variable **V2** (of sustainability) = *Technique (available and necessary)*

Activity *a2* (in service of sustainability) = Competitive Engineering *(Value Analysis)*

Variable **V3** (of regeneration) = *Knowledge (informational primacy)*

Activity *a3* (in service of regeneration) = *Prototype (CAD, ISO), including Spin-offs*

For Figure 5.7, we have MODULE-6 of scientific entrepreneurship with knowledge transfer:

TYPE OF BUSINESS = *outsourcing*

MODE/SCIENTIFIC ENTREPRENEURSHIP MGM = *Cooperation and networking*

Scientific management function = *Logistic Chain Availability*

Variable **V1** (of competitiveness) = *Resources (available and necessary)*

Activity *a1* (in service of competitiveness) = Utility *(of knowledge transfer)*

Variable **V2** (of sustainability) = *Knowledge (plus-knowledge)*

Activity *a2* (in service of sustainability) = *Groupware techniques (communication, collaboration, coordination)*

Variable **V3** (of regeneration) = *Technical (permanent upgrading)*

Activity *a3* (in service of regeneration) = *Team Building (adjusting interorganizational culture)*

For Figure 5.8, we have MODULE-7 of scientific entrepreneurship with knowledge transfer:

TYPE OF BUSINESS = *polemocracy*

MODE/SCIENTIFIC ENTREPRENEURSHIP MGM = *Marketing management (marketing intelligence)*

Scientific management function = *Indexation-measurement*

Variable **V1** (of competitiveness) = *Knowledge*

Activity *a1* (in service of competitiveness) = *Geo-marketing (including political marketing)*

Variable **V2** (of sustainability) = *Utility*

Activity *a2* (in service of sustainability) = *Technological Alert ("week signs" in the market)*

Variable **V3** (of regeneration) = *Resources (including their diversification)*

Activity *a3* (in service of regeneration) = *Relationships with clients/those interested (post service/product).*

For Figure 5.9, we have MODULE-8 of scientific entrepreneurship with knowledge transfer:

TYPE OF BUSINESS = *animal spirits*

MODE/SCIENTIFIC ENTREPRENEURSHIP MGM = *Knowledge management*

Scientific management function = *Knowledge audit*

Variable **V1** (of competitiveness) = *Indexation-measurement (identifying newly entered, substitute products, etc.)*

Activity *a1* (in service of competitiveness) = *Knowledge map (permanently updated and expanded)*

Variable **V2** (of sustainability) = *Availability (including political and governmental international lobby*

Activity *a2* (in service of sustainability) = *Management of Intellectual Property (protection of data)*

Variable **V3** (of regeneration) = *Documentation (continuous learning)*

Activity *a3* (in service of regeneration) = *Management of Documents (archiving, coding)*

There are two aspects to be noticed, a general one and a specific one, at the end of the present methodological analysis:

- We can notice how the traditional functions of modern age management (mass, industrial planning) are broadened and diversified into new forms of postmodern management, with functions corresponding to the ones of the strategic cube of research–development–innovation, which might suggest some sort of "scientific management."
- The present patterns/models (i.e., the management through interface/by module 5, knowledge management/by module 8, marketing management/by module 7, and predesign management/by module 1) correspond to models/modes 0, 1, 2, and 3 proposed by Zalewska-Kurek et al. (2016), as shown in the section Cognitive Analysis and Table 5.1, respectively, yet with multiplicity valences, both intensively (through/of/in themselves) and extensively (beyond them).

The models/patterns presented wait for case studies to be validated and data correlated, which would correspond to a *data-driving* approach! However, these later quantitative analysis benefits of the present essay-study of type *question-driving* in the context of postmodern strategic management applied to scientific knowledge transfer to the nanotechnological environment.

5.4 CONCLUSIONS AND PERSPECTIVES

What is a dispositive (device)? This is the question of one of the most important contemporary Italian philosophers, Giorgio Agamben (2012), which brings us back to the main issue in the Introduction to the present study: science versus technology! With an interesting genealogy, the term seems to have its origins in Hegel's work "The positivity of Christian religion" (*Die Positivität der christliche Religion*); then, taken over by Jean Hyppolite in his "Introduction to the Philospy of Hegel's History," which, by the end of the 60s, inspired Michel Foucault (a postmodernist/poststructural philosopher) to adopt the term "dis-positive." It eventually occupies some strategic positions in a network and replaces the Hegelian universals, materializing them (as in passage, by projection, it supports the fall from sacred to profane) in juridical, military, and technological triptych: LAW (namely, *dispositio*)—MECHANISM (as a process, namely, *Ge-stell* = display and ~*Gerät* = machine)—POWER (by control, ruling—namely, *oikonomia* – including divine order, since in certain Gnostic sects "*ho anthropos tes oikonomia*" = "economy's man" was the embodiment of Christ sign as He has the sacrality in His "roots"). All of them are three derived forms, apparently correlated, but each on "its own direction of manifestation" generating the technical universe of devices with various developmental ability (*Entwicklungsfähigkeit*, in Feuerbach's philosophy).

In such tripartite context, the devices are three times defined as: (1) the whole of erogenous things (linguistically asymmetrical and nonlinguistic—namely, logos and its embodiment) placed into a network—sometimes the network itself as well, thus of substantial nature, (2) having a strategic function, thus procedural, and (3) placed at the intersection of power relations (namely, the need for autonomy) and knowledge relations (namely, the need for strategic independence). The passage of the dispositive from sacred (as an attribute and belonging to the sphere of deity, immutability) to profane is made through *con-secretion,* which corresponds to "restitution of the use and *property to the people"* (according to Trebatius); accordingly, from the "second sphere" appears the dispositive (as a network) populated by instruments, objects, gadgets, devices—or in one phrase—"as technologies of all kinds."

Such opening toward social use—that is, by masses use—gets the separation of the self (namely, inner self, sacred) from the subjective ego (con-secreted, passed to profane), so with the price of losing freedom exactly by accepting "entering in the dispositive" a.k.a of "strategic positioning in the network." From now on, the ruling (oikonomia) is the attribute of the network itself!

Thus we arrive at the postindustrial democracies (bloom), "of man improved by/with technical ideologies" (Dâncu 2016): *affective computing* namely, Afdex technology and Afectiva program; instruments as Mood Scope and Map pines of Quantified Self association—quantifying happiness and its dynamics; *technological singularity* (namely, the dual of the cosmic dynamics related to the genesis of the universe) with the promise, in a Messianic way, of the super intelligence based on technological acceleration and innovation; going beyond the biological barriers and its limitations—namely, chronic, metabolic, oncogene diseases—by shaping the bionic, the infosphere, which eventually induces the *cognitive overflow syndrome* (COS); all this leading to the "growth of human insignificance," so reducing him to a share in a dispositive/network, possibly with the option to confirm with a "like," too often similarly to a "good-bye" from the lost paradise of things "in themselves."

Under these circumstances, is there any chance left for reconciliation of the sacred and the profane, science and technology?; to recover the original sacredness the "primer" as Blaga used to call it? Is there the possibility of "a second curve" (Handy, 2016) without totally compromising humanity enrolled in the device and the economy reduced to power? Fortunately, the answer is: YES! In fact, the present study proposes as "the second curve"—"the second cube." It is actually the double cube which has inscribed in itself another cube, which occupies the inner tetrahedral interstices of the former, (the big one). Through reciprocal projections, it provides the nominative-denominative flow, sacred and profane, science and technology, idea-innovation—all under what was called as the scientific/academic/research entrepreneurship. The scientific knowledge-transfer actions from the sphere of fundamental science into the technological one corresponds to the "golden" dynamics (management) which ensures a fragile balance (in the sense of informational asymmetry between scientific and technical, economic, business infocubes here), but nevertheless competitive, sustainable, and regenerative—the strategic diamond (Fig. 5.10).

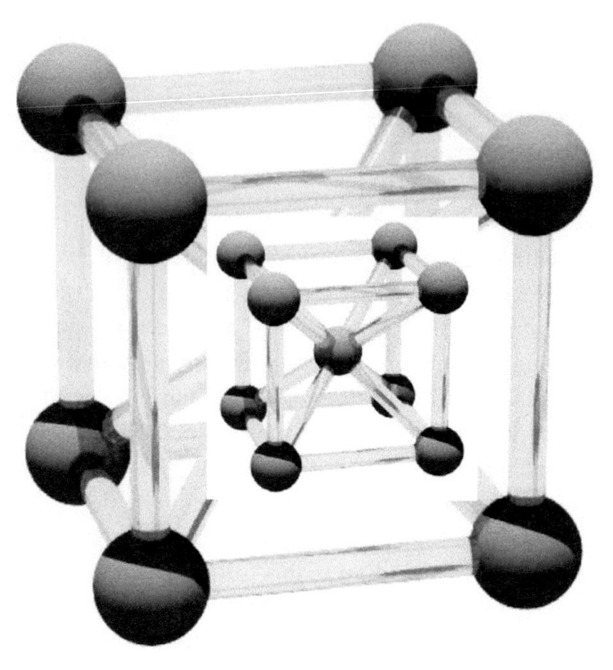

FIGURE 5.10 Strategic diamond: cube in the cube or "the second cube."

At the infinite horizon: science and technology…sacred and profane of postmodern times! And again Agamben (2012): "*The problem of devices profanation* [namely, their being embedded in a background sound, as *blancheur* (Le Breton 1999) of post-modern world.]—*respectively, the problem of restitution for common use of what had been put together and separate them* [namely, exceeding connectivity even totally, but lacking the consistencies/relational intensity; a.n.]—*it is even more urgent* [namely, it can turn into a global–local—glocal problem/illness]. There will be impossible to approach it correctly if those in charge are not able to act upon subjectivity [just like above, a.n.] and also on dispositive. The ultimate goal should be to bring to surface that Ungovernable [what transcends any form of management] *which is the beginning* [Science as sacred foundation of humanity, its immovable property, its common, collective conscience, its latent potential manifested through progress] *but, as well, the vanishing point of any policy* [namely, Platonic desiderate of the Republic governed by philosophers and in the postmodern era—by the authentic scientific eminences, or by we may call—as the author of the present study likes to state—the moral power of value]."

KEYWORDS

- **science and technology**
- **strategic management**
- **strategic cube**
- **scientific entrepreneurship models**
- **knowledge-transfer activities**
- **sacred and profane**

REFERENCES

Agamben, G. *What Is a Dispositive? [in Original Che cos'è un dispositivo?, Nottetempo srl, 2006]. Essay from the Volume: The Friend and Other Essays (Read in Romanian as: Prietenul şi alte eseuri*; Humanitas Publishing House: Bucharest, 2012.

Bekkers, R.; Bodas Freitas, I. M. Analysing Knowledge Transfer Channels Between Universities and Industry: To What Degree do Sectors Also Matter? *Res. Policy* **2008,** *37* (10), 1837–1853.

Carbon Topology. 2013, http://www.ccsem.infn.it/ef/emfcsc2013/pdf/Summary2013.pdf

Dâncu, V. S. *These Wonderful Machines and their Imperfect Slaves [Originally in Romanian as: "Aceste maşini minunate şi scalvii lor imperfecţi!"] Essay in the Volume Politically Incorrect – Strategies for a Possible Romania. Transylvania School [Şcoala Ardeleană, in Romanian Original]*; Publishing House, Cluj-Napoca: Romania, 2016.

Elkington, J. *The Chrysalis Economy – How Citizen, CEOs and Corporations can Fuse Values and Value Creation*; Capstone, John Wiley & Sons Co.: Oxford, UK, 2001.

Gibbons, M.; Limoges, C.; Nowotny, H.; Schwartzmann, S.; Scott, P.; Trow, M. *The New Production of Knowledge. The Dynamics of Science and Research in Contemporary Societies*; Sage Publications: London, Stockholm, 1994.

Gueth, A. Entering into an Alliance with Big Pharma. *Pharm. Technol.* **2001,** October, 132–138.

Gummesson, E. *Marketing Relazionale – Gestione del marketing nei network di relazioni* [în Original *Total Relationship Marketing*, 1999, 2002]; Ulrico Hoepli Editore Spa: Milano, 2006.

Handy, C. *The Second Curve. Thorughts on Reinventing Society*, Random House Business, 2015 [Read in Romanian as *A doua Curbă–să gândim diferit despre viitor*]; Publica Publishing House: Bucharest, 2016.

Haspeslagh, P. C.; Jemison, D. B. *Managing Acquisitions: Creating Value Through Corporate Renewal*; The Free Press: New York, 1991.

Kuhn, T. S. Historical Structure of Scientific Discovery. *Science* **1962,** *136*, 760–764.

Kuhn, T. S. *Historical Structure of Scientific Revolutions*, 2nd ed.; University of Chicago Press: Chicago, 1970; p 55.

Le Breton, D. *L'adieu au corps*; Métailié: Paris, 1999; p 44.

Lyhne, I. *Strategic Environmental Assessment and the Danish Energy Sector: Exploring Non-programmed Strategic Decisions*; Institut for Samfundsudvikling og Planlægning, Aalborg Universitet: Aalborg, 2012.

Perkmann, M.; Walsh, K. University-industry Relationships and Open Innovation: Towards a Research Agenda. *Int. J. Manag. Rev.* **2007,** *9*, 259–280.

Perkmann, M.; Tartari, V.; McKelvey, M.; Autio, E.; Brostrom, A.; D'Este, P. et al. Academic Engagement and Commercialization: A Review of the Literature on University-industry Relations. *Res. Policy* **2013,** *42*, 423–442.

Putz, M. V. The Strategically Dynamics of the Research–Development–Innovation Potential in the Nano-Technological (Meta)clusters [Originally in Romanian as: Dinamica strategică a potentialului de cercetare-dezvoltare-inovare in (meta)clustere nanotehnologice.] PhD Thesis, Faculty of Economics and Business Administration of West University of Timisoara, in Preparation, 2016–2019.

Putz, M.V. Strategic Cube of the Organization Competitive Advantage. [Originally in Romanian: Cubul strategic al avantajului competitiv al organizațiilor]. MBA Thesis, Faculty of Economy Science and Business Administration, West University of Timişoara, 2017.

Putz, M.V. *The Strategic Cube of the Distinctive Advantage. Epistemological Approach*, Chapter 2 of the Present Monograph, 2019.

Rothaermel, F. T.; Agung, S. D.; Jiang, L. University Entrepreneurship: A Taxonomy of the Literature. *Ind. Corp. Change* **2007,** *16* (4), 691–791.

The Foundation and the Center for the Scientific Culture "*Ettore Majorana*", 2013, http://www.ccsem.infn.it/.

The Pope John-Paul the 2nd, 1993, http://w2.vatican.va/content/john-paul-ii/it/speeches/1993/may/documents/hf_jp-ii_spe_19930508_scienziati-erice.html

van Rijnsoever, F. J.; Hessels, L. K.; Vandeberg, R. L. J. A Resource-based View on the Interactions of University Researchers. *Res. Policy* **2008,** *37*, 1255–1266.

WFS, The World Federation of Scientists, 2013, http://www.federationofscientists.org/

Zalewska-Kurek, K.; Geurts, P. A. T. M.; Roosendaal, H. E. The Impact of the Autonomy and Interdependence of Individual Researchers on Their Production of Knowledge and Its Impact: An Empirical Study of a Nanotechnology Institute. *Res. Eval.* **2010,** *19* (3), 217–225.

Zalewska-Kurek, K.; Egedova, K.; Geurts, P. A. T. M.; Roosendaal, H. E. Knowledge Transfer Activities of Scientists in Nanotechnology. *J. Technol. Transf.* **2016,** doi:10.1007/s10961-016-9467-6.

Ziman, J. *Reliable Knowledge: An Exploration of the Grounds for Belief in Science*; Cambridge: University Press, 1991.

CHAPTER 6

Business Strategies by the Multinodal Logistics Within the Cubic Network of Distinctive Advantage

ABSTRACT

In the new-economy contemporary context, one can find (in fact rediscover) the five business strategies (i.e., the hierarchic, of proximities, of networking, of cube, and of deep learning strategies, respectively), with different implementing-operational potentials. In different shapes (paradigms) they transform the business chains in a multiple (and finally multinodal) logic (logistic) forms. As a result, they sequentially unify the business behavior from the "red ocean" perspective through competition toward the wise business, with the aid of synergic competitive, sustainable, and regenerative advantage.

Motto:

"Nathan: 'This isn't a house, it's a research facility.'"

—*Ex-Machina* (2015)

© Universal Pictures International & Co.

6.1 INTRODUCTION

One of the strangest phenomena of the postmodern era of "*the new economy*" is "*the opening that closes*": manifested mainly in Japan, on a social level and known as *Hikikomori* (ひきこもり or 引き籠り)—it describes the reluctant withdrawal from the social life, for at least 6 consecutive months, without an apparent reason or as a consequence of a metabolic or chronic disease. It is a form of "postmodern hermitism" through which the individual, especially teenagers, physically isolate themselves from the world—though they are hyper-connected to it through the virtual environment via computers (which, nevertheless, cancels the idea of zero experience, as sometimes is wrinkly assumed). On the level of organizations and business environment, it is manifested through complex competition and collaboration strategies, co-petition, fierce fighting for resources, investment in the "red ocean," *lose–lose* strategies (animal spirits), in parallel business, and in different chains of values. Thus, the postmodern way in organizing logistics is assigned with a diversified role in the strategic management: not only to provide, on the operational and tactical levels, making business plans designed in the business mission and objectives but also in identifying the "*optimal path,*" potentially minimized. The organization and its business transform toward competitive, sustainable, and regenerative advantages, by diversifying the portfolio of goods and services and by adjusting—alignment to current technology in the development of communication, transport, and trading. In this context, the postmodern logistics has the role of "dynamic profiler" of the business area, embedding, in a first ranking stage (mainly for large societies, the bureaucratic ones) chains of value and supplying chains with feedback loops, but also with superior extensions, while passing from efficiency to tactics, to strategy (Fig. 6.1), in a variety of manifestations, as it is further illustrated:

- the chain of cyclic value as a pattern with sustainable potential in planned economy (Govindan et al., 2015);
- management of the chain of value (supply of goods and services) in a collaborative manner, potentially integrated and integral, with application to supply by sea (Ascencio et al., 2014), respectively, in the logistics industry in its widest meaning (Pateman et al., 2016);
- the expansion of the multidimensional logistic network, with potential in military industry strategies and in military organizations/organization strategies (Jie and Wen, 2012);
- sustainable logistics channels (environment, by recycling products, services, and waste) in econo-ecological strategic management of multiple stakeholders for sustainable businesses in econosphere (companies, consumers, government, and environment itself; Fujimoto, 2012);
- the management of the supply chain with integration of the concept of humanitarian technology is mainly in relation to recycling of the food products in food units, restaurants, and hotels; this is

possible by adapting the technological support in satisfying the social needs for the purposes of eradicating poverty—along with the whole value chain for food products and services (Crumbly and Carter, 2015);

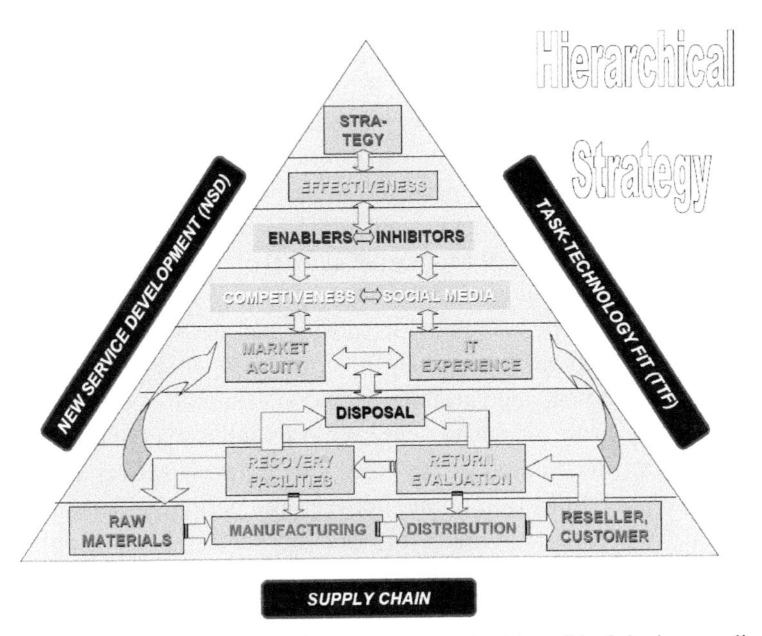

FIGURE 6.1 Multilayered strategic management, in hierarchical logics, applied in automobile industry, targeting adaptability to the market fluctuations regarding demand in diversity, and load, by maintaining competiveness in conditions of change in the social and economic factors.

Source: Adapted with permission from Küber et al. (2016).

- improving the companies' services by adopting a viable logistic pattern (even if variable), by a holistic, systemic approach, between the supply needs, the supply chain, and the supply technique (Stich and Groten, 2015);
- formulating a multi-criteria dynamic model, synergic, for transport units with exploitation of organization-technological structures for railway transport organizations between the company divisions and opportunities/constraints of internal/external factors of transport process, including additional packaging, labeling, sorting, design, cargo-insurance functions/services, by optimizing traffic options (Lomotko, 2016);

- a modular approach of distribution logistics, from planning to the procedural course of goods and services, with multiple functions, with a client/service-oriented architecture, through an integrative management, dynamically inventorying business options (Kamphues and Hegmanns, 2015);
- approaching the risk by the logistics of externalized units/functions (by *outsourcing logistics*); it has the potential of optimizing the costs and enhancing the service quality by systematical exploitation (potentially dynamic) of capabilities, under the circumstances of minimizing the risks; the noticeable case is that of pharmaceutical industry simultaneously connected to design, resources, production, and supply (El Mokrini et al., 2016).

The "bureaucratic" strategy as positioned in postmodern era, and mainly because of that, is merely a "container" of strategic, historical, generalized, adapted, and variable variations. It should open the integrated strategy of the circle, which however, recloses at the level of organizational vicinity, by various functions and roles it assumes. However, the logistics of the emerging "inflexion point" between the peripheral zone of change (external, exogenous environment) and the central zone of culture and institutional rites (internal, endogenous environment) is to be analyzed in the next section.

6.2 COGNITIVE ANALYSIS

The second level of "opening that closes" strategy is the proximity one, oriented and applicable mainly in the "smart" urban orientation of postmodern logistics (Fig. 6.2):

- solving humanitarian emergencies (urban disasters, earthquakes, and fires) by mixed, probabilistic—stochastic modeling of uncertainty, by using a logistical structure of proximity, namely, local distribution centers around a data center (*warehouse*) dynamic monitoring of logistic problems, including the post-disaster phases (Tofighi et al., 2016);
- the use of multimodal logistics in designing a smart transportation system, even in the absence of the communication and informatics technology, is based on previously received messages; it puts into

practice the dynamic principle of "*track and trace*" in order to optimize the dynamic capacities with the high economic impact, to minimize repercussions; it has a vital applicability in maritime transport (Mondragon et al., 2012);

FIGURE 6.2 Multilayer strategic management in the econosphere: the logic of inter- and overlayer proximities features inventory strategic logistics (as a balance) between the exogenous manifestations (centrifugal) related to economic efficiency, flexibility and management of change—and the endogenous (centripetal) ones of assembly (design and production), transport (communication in a broad sense), and needs (including humanitarian). It has the origin in the organizational cache (software) of society (business, system, city, and nation) as acting in the available business space.

- the identification of key factors (e.g., the strategy of operations, infrastructure logistics, innovation and ideas, marketing, people/ human resource regulations, the laws of environment/location, and financial resources) in the logistics of smart cities—enables the achievement of aspirations; it enhances the mobility of sustainability and the quality of life in urban areas with the allied beneficiaries of complex economic activity; the multi-stakeholders, all being cross-connected and eventually integrated (Kiba-Janiak, 2016);

- cocreation of value through logistics of smart cities—through identifying the profile (pattern) of endogenous processes, basic (in-store processes) as well as the barrier (of potential) between the operational core of endogenous activities (in the city) and proximity (superimposed proximity) of services (Gammelgaard, 2016);
- innovation in online trading may have limitations in areas of proximity of goods and services, particularly in food, supermarkets, and grocery stores, but can be both catalyzed by a specific logistics concept and specifically adapted to the culture of the area (region, country), see Saskia et al. (2016);
- Complementarities between urban logistics and the physical Internet (possibly on street by wireless) can be exploited in a synergic sense, by producing the so-called hyper-connected city logistics; thus, that city changes into a hub (center) facilitating the communication in the global network (WWW, World Wide Web). It provides a standardized benchmark, within a multilayered activity space, yet within an open space of knowledge (namely, open access), with a modular container of goods and services, including the means of transportation/relocation/connection for freight and people; as a consequence, a dynamic map for effective, controlled, and sustainable regional development may effectively be produced (Crainic and Montreuil, 2016);
- the use of green solutions proves positively correlated, although very weakly, with the availability of generation Y (media society); they are eager to pay more for logistics oriented to ecological environment, and uncorrelated in fact with Millennials. This behavior indicates a potential danger (by passivity, isolation, indifference, and even arrogance) to urban problems of a society with high density as in the large cities. From this, the need for mixed methods of marketing–logistics in developing awareness of the advantages of econo-ecology through sustainable business (Moroz and Polkowski, 2016).

Yet the question arises: "Is it possible, a strategy that closes with the opportunity of extension? Compact filling of business space by intelligent joining of any node-organization in a network/meta-structure/cluster with any other, on discrete paths (transformational), cyclic and optimal?" The answer is positive, in the framework of the strategy of distinct advantage cube, see the next section.

6.3 METHODOLOGICAL ANALYSIS

Currently, a particular type of business strategy (Putz, 2017, 2019) has been identified, toward transforming the launching on a market with fierce competition ("Red Ocean") to an eco-ecological integrated business, distinctive, triple-fold wised: with competitive effect (relating the resources), sustainable on a social level, and regenerative (economically) even if with reduced profits, say in a "blue ocean"; it features so the continuous potential to increase, relaunch, reinvent, recycle, reorient in the (complex) space of business development (from entrepreneurship to maturity) and on the second curve, and so on. It also has a feature of being a discrete transformation not just incremental, yet not necessarily disruptive either, but rather promoting "the close-to-close" modeling strategies and the dynamic business model, self-transformational and ultimately transforming society. The cube of distinctive advantages shifts economy to a smart balance with the ecological environment, designing network strategy of the logistics of smart city (valid for meta-hierarchical organizations, and cluster structures as well), without leaps, information gaps, and planning management. This model provides both, closure (intra-organizational strategy), and opening (exogenous strategy, expansion of business space, transport, and communication of economic information, and "beyond" business itself)—all through the uni-, bi-, and multinodal opening logistics, development, and welfare. Under these conditions:

- Figure 6.3 illustrates "specific switching" from proximity organization to organized—meshed but still 2D, in perspective of the strategic cube, but still maintained within a strategy as neural network, here particularly with the distinct advantage of urban areas, according to Porter (1995), and Kasinitz and Rosenberg (1993);
- Figure 6.4 illustrates the strategic cube of competitive, sustainable, and regenerative advantage (Putz, 2017, 2019), as an "ordered lift in space" of the 2D projection neural network of Figure 6.3. Thereby, one is achieving expanded predominant type of business within a particular node of the cube, and respectively, united (from the most unsustainable business, the state VIII of animal spirits to wise business in the state I). Along the "Hamiltonian path" one may "reach" every nod/type of business—only once. With the endogenous abstract logistics of the Hamiltonian path/chain, one "makes sense" in crossing the cube from one business to another,

in relation to the positive directions of achieving competitiveness (+0X), sustainability (+0Y), and regenerative (+0Z) attributes;

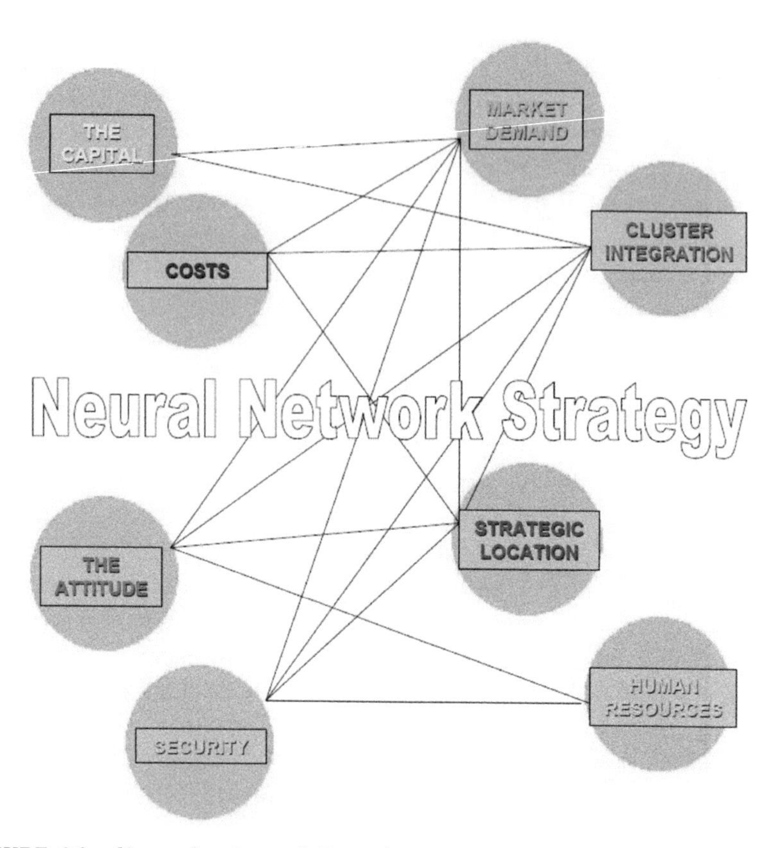

FIGURE 6.3 (See color insert.) Strategic management of networking as a linkage (not necessarily with networking consistency) in the multinodal logic (logistics); note the catalyst factors (as integration in conglomerates–clusters, the market demand, the human resources, and the strategic placement) along the constraint factors (as the cost, the capital security, and the strategic attitude) in the space of business development; here generalized according to Porter (1995).

- Figures 6.5–6.7 expand the strategic cube of competitive, sustainable, and regenerative advantage "beyond it," by exogenous, mono-nodes interaction with the same Hamiltonian path and length of crossing businesses; one gets a superior logistics by reducing "logical distance" to values "+1" in the optimized case, and to "−1" in the minimized case (Fig. 6.7), both in relation to the endogenous value of the logistic chain of transforming businesses ("+3") of Figure 6.4;

- Figure 6.8: the case of the strategic cube of competitive, sustainable, and regenerative advantage in bi-nodal exogenous interaction with the same Hamiltonian path as in the previous cases, but with a minimal logistic chain "−1" as in case illustrated in Figure 6.7;

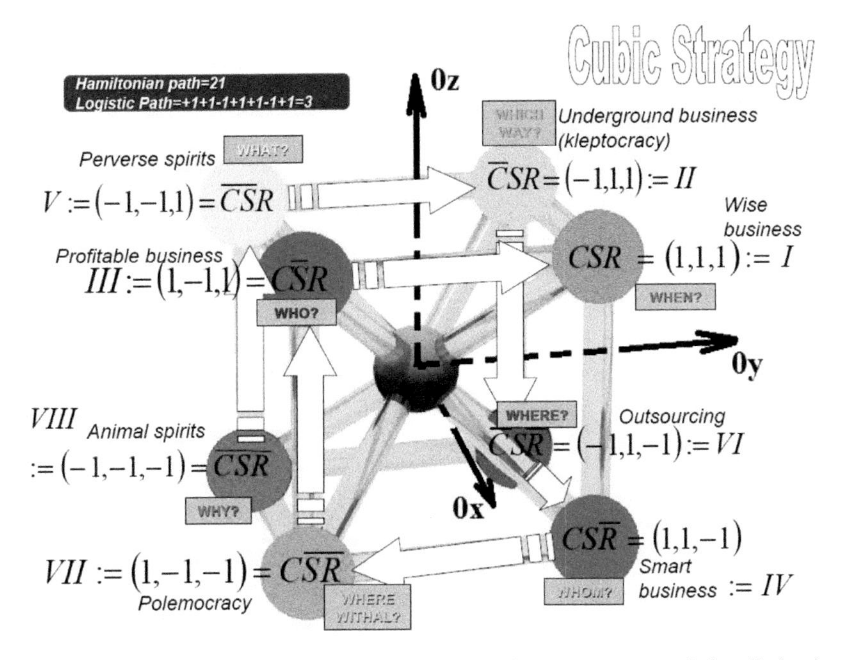

FIGURE 6.4 (See color insert.) The 3D strategic management of the distinctive advantage cube of the competitive, sustainable, and regenerative advantage, as developed with the multinodal course of dual quantification: by (1) the longest path, being self-referential, along the so-called Hamiltonian path; and by (2) the logistic path, exo-referential, and positively oriented to satisfying/reaching the competitive (on 0X), sustainable (on 0Y), and regenerative (on 0Z) advantage. It nevertheless gives consistency (or effective relationship) to 2D connectivity of the network of the flat strategy of Figure 6.3.

- Figure 6.9: the strategic cube of competitive, sustainable, and regenerative advantage, in exogenous-multinodal interaction, with the same Hamiltonian path as in the previous cases, but with an optimal logistic chain "+1," as in cases in Figures 6.5 and 6.6.

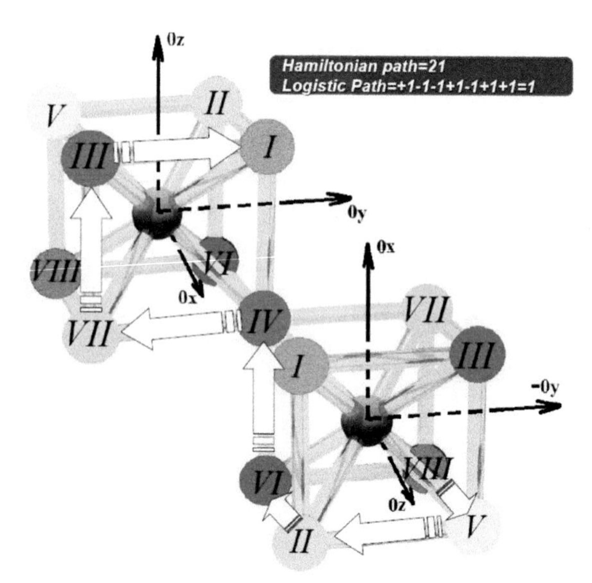

FIGURE 6.5 (See color insert.) Uni-nodal interaction (by a "common tip") of the strategic cubes in order to achieve a competitive, sustainable, and regenerative business, with the same Hamiltonian path as in the endogenous cube of Figure 6.4, but with an optimized logistic path.

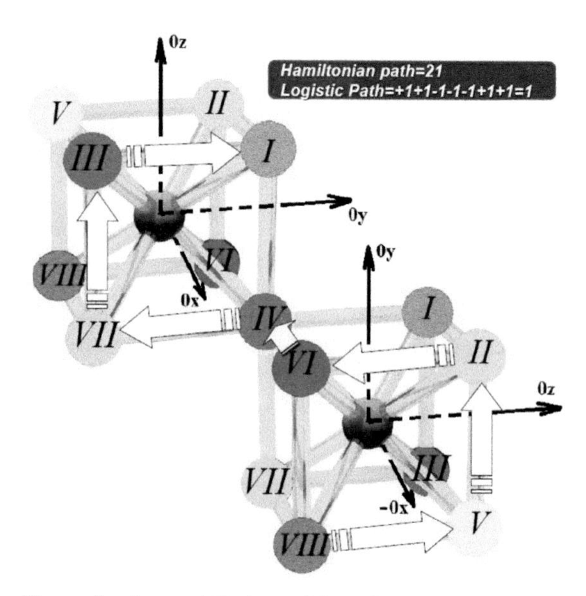

FIGURE 6.6 (See color insert.) Variant of the uni-nodal interaction of Figure 6.5.

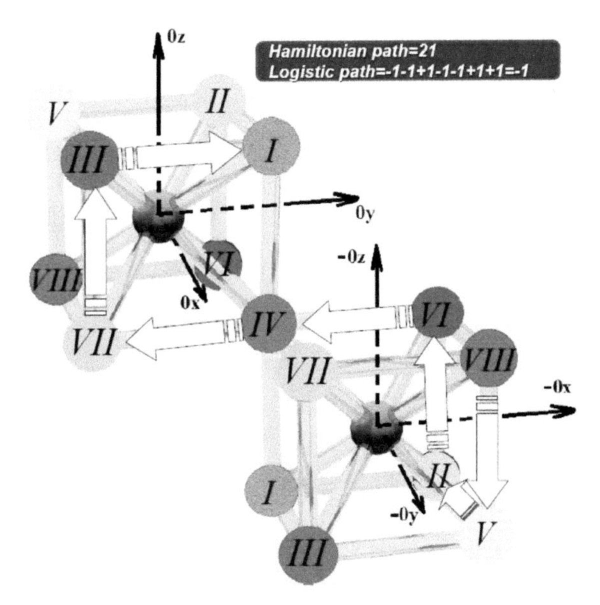

FIGURE 6.7 (See color insert.) Variant of the uni-nodal interaction of Figure 6.5, but with a minimized logistic path.

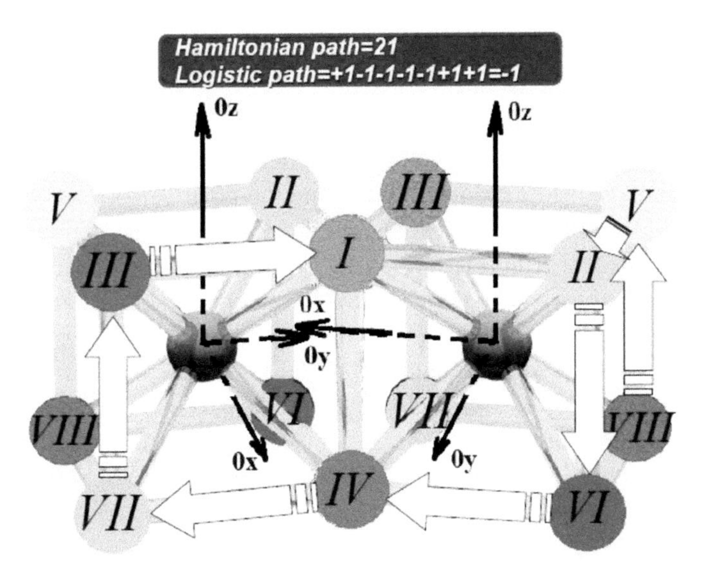

FIGURE 6.8 (See color insert.) Bi-nodal interaction (by a "common side") of endogenous strategic cubes of Figure 6.4, with the same Hamiltonian path, but with a minimized logistic path—at the value in Figure 6.7.

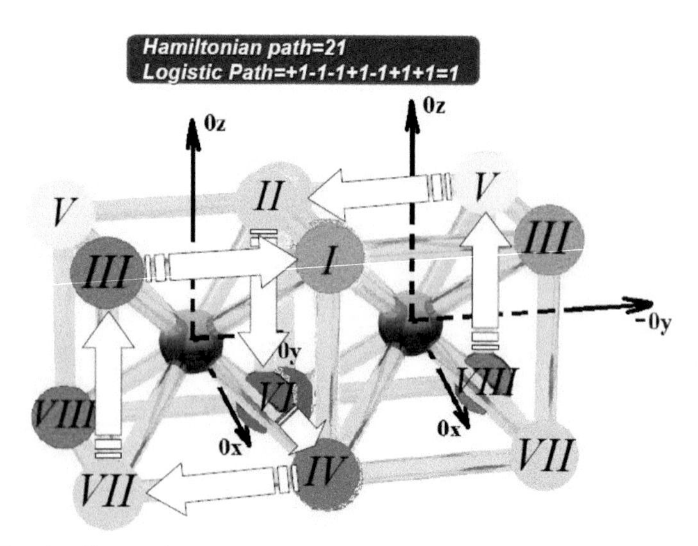

FIGURE 6.9 (See color insert.) Bi-nodal interaction (by a "common side") of endogenous strategic cubes of Figure 6.4, with the same Hamiltonian path, but with an optimized logistic path—at the value in Figures 6.5 and 6.6.

The result is bi-nodal, exogenous, with the cubic logistic in Figure 6.8 as being the most adequate configuration able to minimize the logistic chain, with "the most numerous nodes—fix points." Note the common axis of wise and smart businesses, thus making both businesses profitable and legal (in the "smart" node, IV), but also sustainable (in the "wise" node, I) in the cubic business space modeled by the cube of the distinctive advantage. More than demonstrating the positive correlation between multinodal logistics and the exogenous opening of the distinctive advantage cube, the present result, regarding the minimization of the logistic chain in the exogenous bi-nodal cubic development of the successful business (i.e., which manages to pass from the "red ocean" to the "blue ocean"), opens the way for a new type of strategy, that is, the *post-networking strategy* (Putz, 2019; Putz and Petrişor, 2020). The management strategy is a continuous profound learning for developing and transforming (the life cycle) a business, see also the forthcoming section for conclusions.

6.4 CONCLUSIONS AND PERSPECTIVES

From *Hikikomori (isolation)* to *Keikaku (planning)*: the postmodern logistics "overthrows" and combines the two in an open–closed cycle of goods, services, a.k.a. "particle–wave" in the new economy, quantum economy! Logistics in the network, by combining logistic chains in logistic networks, multimodal, sustainable eventually, requires complex modeling and analysis: its methods vary from algorithms for linear mixed programing, by what may be called as the intelligent calculation of production engineering (Confessore et al., 2013), to integrated processes logistic, by using the framework of artificial intelligence, in general, and of machines with adaptive learning (*learning machines*) in particular (Knoll et al., 2016).

Thus, production is virtually planned, distributed, promoted, sold, recycled, reintegrated, and a priori (pre)adapted to the ideal model of competitive, sustainable, and regenerative business by the multiple quick-change reconfigurable (agile–changeable–reconfigurable) characteristic. Perhaps not by chance the drug pharmaceutical business moves jointly with the market developments, including risks, fluctuations, and its profitability. The drug design is the domain of frontier in natural sciences (combining nanochemistry with nanobiology), which use most intensely learning, by the adaptive-machine algorithms and databases. Processing the myriad of data (properties) of molecular fragments aims in their successful combination, in order to generate an active potential (in an active substance thus designed) with beneficial effects. However, note that sometimes beneficial means detrimental—especially in fighting geno-toxic, through and for the induced cellular apoptosis in the body/organ/receptor's bio-molecules (Wilson, 2017).

Similarly, adaptive, quantum economy, based on nanotechnologies, and complex algorithms for analysis and prediction, reaches the age of adaptive learning, through the algorithms and multinodal procedures (logistics). However, the question—can such an economic approach be integrated in a coherently conceived model, even disruptive but with pre-established origins?—has given a positive answer by the present study. Adaptive learning, anticipating included configuration, has its origins in modeling modal networks (network-modeling), which already corresponds to postmodern approach of economy in the sense of connectivity and networking resources and production factors (labor, capital, and management), marketing, and global impact of innovation (even local).

The business network approach has its origins in the structured variants of proximity approach type (intra- and interlayers) in organization, planning—logistics and marketing. It is also a "democratic" form of modern facets of open economy based on the logistics chain, while "bureaucratically" supported by technological services, as synthesized in the present study, (Fig. 6.1–6.3). However, the shift from the postmodern approach of networking to *post-economy* (we may say) may be performed by using the strategic cube of competitive, sustainable, and regenerative advantage. It features the key strategy in exogenous opening of "paths" and connections, transforming the business space into a logistic chain of transformations of a business. The resultant is an expanded network with mono-, bi-, and multinodal interaction in the economy space covered by the strategic cube of competitive, sustainable, and regenerative advantage, namely, Figure 6.10. Thus, the present study illustrates as well *(proof concept)* the way in which multinodal logistics in the exogenous strategic cube of distinctive advantage corresponds to a sustainable transformation of the GLOCAL business space (global–local).

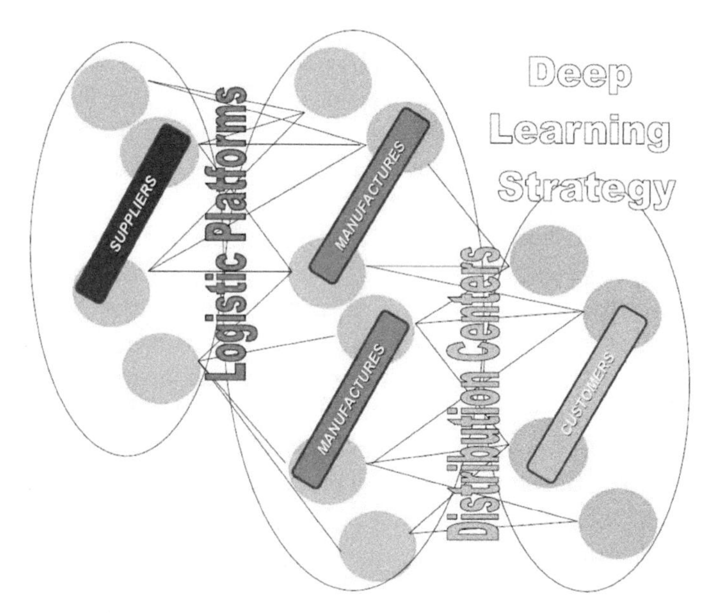

FIGURE 6.10 (See color insert.) The multinodal strategic management of profound learning as an extended version (exogenous by "beyond itself") through the (self) interactive networking strategy of Figure 6.3; we can notice the "producers' layer" interposed to the specific layers for suppliers and customers, through interfaces containing the logistic layer of intercommunication and (re)distribution of value (goods and services); this way the feedback and a dynamic filter is providing, along the multilayered, and learning complexity. The model helps in formulating predictions and advancing proactive attitudes in developing competitive, sustainable, and regenerative businesses, with a depth learning architecture (recycled), namely, deep learning machine, passing from artificial to natural and regaining the sustainable (econo-ecolo) business space.

KEYWORDS

- **strategic management**
- **hierarchy (decisional tree)**
- **vicinity culture**
- **networking logic**
- **strategic cube**
- **coupled (uni-, bi-, and multinodal) strategies**
- **strategic teaching algorithms**

REFERENCES

Ascencio, L. M.; González-Ramírez, R. G.; Bearzotti, L. A.; Smith, N. R.; Camacho-Vallejo, J. F. A Collaborative Supply Chain Management System for a Maritime Port Logistics Chain. *J. Appl. Res. Technol.* **2014,** *12*, 444–458.

Confessore, G.; Galiano, G.; Liotta, G.; Stecca, G. A Production and Logistics Network Model with Multimodal and Sustainability Considerations. *Procedia CIRP* **2013,** *12*, 342–347.

Crainic, T. G.; Montreuil, B. Physical Internet Enabled Hyperconnected City Logistics. *Transp. Res. Procedia* **2016,** *12*, 383–398.

Crumbly, J.; Carter, L. Social Media and Humanitarian Logistics: The Impact of Task Technology Fit on New Service Development. *Procedia Eng.* **2015,** *107*, 412–416.

El Mokrini, A.; Dafaoui, E.; Berrado, A.; El Mhamedi, A. An Approach to Risk Assessment for Outsourcing Logistics: Case of Pharmaceutical Industry. *IFAC-PapersOnLine* **2016,** *49* (12), 1239–1244.

Fujimoto, H. Strategic Management for Environmental Logistics Channel. *Procedia: Soc. Behav. Sci.* **2012,** *58*, 1443–1447.

Gammelgaard, B.; Andersen, C. B. G.; Aastrup, J. Value Co-creation in the Interface Between City Logistics Provider and In-store Processes. *Transp. Res. Procedia* **2016,** *12*, 787–799.

Govindan, K.; Soleimani, H.; Kannan, D. Reverse Logistics and Closed-loop Supply Chain: A Comprehensive Review to Explore the Future. *Eur. J. Oper. Res.* **2015,** *240*, 603–626.

Jie, W.; Wen, W. Research on 6R Military Logistics Network. *Phys. Procedia* **2012,** *33*, 678–684.

Kamphues, J.; Hegmanns, T. A Modular Approach for Integrated Inventory Management in Distribution Logistics. *IFAC-PapersOnLine* **2015,** *48* (3), 1815–1820.

Kasinitz, P.; Rosenberg, J. Why Enterprise Zones Will Not Work: Lessons from a Brooklyn Neighborhood. *City J.* **1993,** Autumn 1993, 63–69.

Kiba-Janiak, M. Key Success Factors for City Logistics from the Perspective of Various Groups of Stakeholders. *Transp. Res. Procedia* **2016,** *12*, 557–569.

Knoll, D.; Prüglmeier, M.; Reinhart, G. Predicting Future Inbound Logistics Processes Using Machine Learning. *Procedia CIRP* **2016,** *52*, 145–150.

Küber, C.; Westkämper, E.; Keller, B.; Jacobi, H.-F. Planning Method for the Design of Flexible as well as Economic Assembly and Logistics Processes in the Automotive Industry. *Procedia CIRP* **2016,** *41*, 556–561.

Lomotko, D. V.; Alyoshinsky, E. S.; Zambrybor, G. G. Methodological Aspect of the Logistics Technologies Formation in Reforming Processes on the Railways. *Transp. Res. Procedia* **2016,** *14*, 2762–2766.

Mondragon, A. E. C.; Lalwani, C. S.; Mondragon, E. S. C.; Mondragon, C. E. C.; Pawar, K. S. Intelligent Transport Systems in Multimodal Logistics: A Case of Role and Contribution Through Wireless Vehicular Networks in a Sea Port Location. *Int. J. Prod. Econ.* **2012,** *137*, 165–175.

Moroz, M.; Polkowski, Z. The Last Mile Issue and Urban Logistics: Choosing Parcel Machines in the Context of the Ecological Attitudes of the Y Generation Consumers Purchasing Online. *Transp. Res. Procedia* **2016,** *16*, 378–393.

Pateman, H.; Cahoon, S.; Chen, S.-L. The Role and Value of Collaboration in the Logistics Industry: An Empirical Study in Australia. *Asian J. Shipp. Logist.* **2016,** *32* (1), 033–040.

Porter, M. E. The Competitive Advantage of the Inner City. *Harvard Business Review* **1995,** May–June: Politics. https://hbr.org/1995/05/the-competitive-advantage-of-the-inner-city.

Putz, M. V. Strategic Cube of the Organization Competitive Advantage. [Originally in Romanian: Cubul strategic al avantajului competitiv al organizațiilor]. MBA Thesis, Faculty of Economy Science and Business Administration, West University of Timişoara, 2017.

Putz, M. V. *The Strategic Cube of the Distinctive Advantage. Epistemological Approach*, Chapter 2 of the Present Monograph, 2019.

Putz, M. V. The Strategically Dynamics of the Research-Development-Innovation Potential in the Nano-Technological (Meta) Clusters [Originally in Romanian as: Dinamica strategica a potentialului de cercetare-dezvoltare-inovare in (meta)clustere nanotehnologice.] Ph.D. Thesis, Faculty of Economics and Business Administration of West University of Timisoara, in Preparation, 2019.

Putz, M. V.; Petrişor, I. *The Code of Strategic Management: From Modern to Post Modern (Meta) Clustering Approach*—in Preparation, 2020.

Saskia, S.; Mareï, N.; Blanquart, C. Innovations in E-grocery and Logistics Solutions for Cities. *Transp. Res. Procedia* **2016,** *12*, 825–835.

Stich, V.; Groten, M. Design and Simulation of a Logistics Distribution Network Applying the Viable System Model (VSM). *Procedia Manuf.* **2015,** *3*, 534–541.

Tofighi, S.; Torabi, S. A.; Mansouri, S. A. Humanitarian Logistics Network Design Under Mixed Uncertainty. *Eur. J. Oper. Res.* **2016,** *250*, 239–250.

Wilson, E. K. Deep Learning to the Rescue. Pharmaceutical Chemists Pin Hopes on New Machine-learning Method for Drug Discovery. *Chem. Eng. News* **2017,** January 23, 29–30.

CHAPTER 7

Risk Management in Nanotechnology Projects Toward Eight-Fold Ws

ABSTRACT

In the project management context, the risk problems in a nanotechnologic project with econo-ecological relevance are presented and analyzed as an eight-fold perspective of cubic strategic (questions) vertices. The presentation contents are original and can constitute at the same time the basis for risk analysis and management in an applicative nanotechnologic project, and also for a further conceptual extension in the postmodern formulation of the strategic management of the synergic competitive, sustainable, and regenerative advantage.

Motto:

"*A pessimist sees the difficulty in every opportunity;*
an optimist sees the opportunity in every difficulty."
—Sir Winston S. Churchill

7.1 INTRODUCTION

In the post-knowledge society, either now or later, the—economic projects should be based on nanotechnology in all their phases: the research, development, knowledge, technological transfer, and sharing value—all to enhance competitive advantages (Porter, 2008). It features durable advantage on a microeconomic level of organizations (firms, companies, and institutions) and sustainable advantage at a macroeconomic level of regional and global businesses (clusters, regions, nations, multinational, and unions).

In this competitive context, for a RDI project/action (research and innovation action/innovation action, RIA/IA) to influence on short-, medium- and long-term economy and society, either local or global, some key features are required for a measurable impact (perceived), depending on the project objectives, namely (Hristova et al., 2014, 2015):

1) to make/produce something *new and meaningful*;
2) to provide *a solution or an answer to a problematic situation,* currently unresolved and having potential on the local/global, societal impact present and future;
3) to *formulate and implement a comprehensive solution to the specialist in the field,* even if it appears that he did not see the proposed solution.

Such innovative research projects are, inherently, subject to risks, mainly of the following types:

1) *Scientific and technical risks*: on the content of the project
2) *Financial risks*: relative to their financial possibilities (cash flow, debts, loans, maturities, banking entities, or partnerships with government, European Union, World Bank, NGOs, local government, regional, investors, etc.) ensuring financial guarantees for the purposes of the project
3) *Administrative risks*: In connection with the circulation of project documentation and human resources access to tangible and intangible capital of the organization, the partnership relationships of the organization, involvement in consortia, knowledge management related to goals and providing deliverables
4) *Risks of intellectual property rights* (*IPR risks*): It is related to access the intangible classified information in the organization or partners (namely, *open source know-how, knowledge transfer*), but also with communication of the project's results in open formats (open access). They can also be converted into patents/market entry and augmented through knowledge sharing by sale and purchase of patents aiming at relaunching business as a joint venture project, or in association of shared values and knowledge within consortia and clusters (conglomerates), and so on.

All these types of risks and their variant subscribe to stages of identification of risks and then to stages of risk management, in a closed cycle of problematization—learning—reproblematization—and so on to ensure a competitive success of the undertaken project (Fig. 7.1).

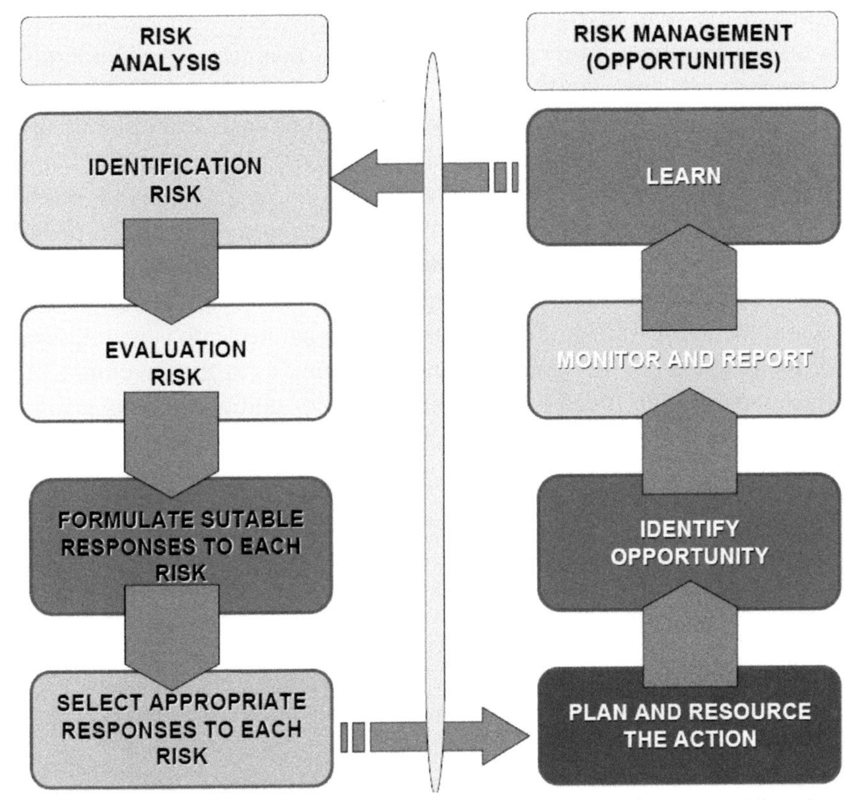

FIGURE 7.1 (See color insert.) Cycle of risk management for successful projects [inspired and generalized according to Hristova et al. (2014, 2015)].

At this point it is worth making the *distinction between risk and uncertainty*:

1) *uncertainty* is associated with unexpected, unpredictable, and can have a negative impact on the project
2) *risk* is associated with the expected, imagined, and therefore faced with the preaction images (scenarios) to overcome them in an operative manner (rapidly), preferably "*in real time*."

Reconciling this duality can be performed by a *tertium datur* method—of the third party included, when we give a chance and prepare the virtual scenario also for unpredictable cases, uncovered by risks. They can be previewed (in a true postmodern logic of the *post-truth*, namely the global word of the year 2016) and respectively, turned into opportunities (with a proper cognitive training), in the inclusive manner of postmodernism. Transforming uncertainty into opportunities depends on creativity and "the art of the project manager," the leader in this case. [S]he should bring out added value of unpredictable risk situations, some benefit, and a real competitive advantage for the project and organization, with positive impact, inclusive, postmodern, postknowledge, and eventually a positive post-truth (in the sense of postrisk postanticipation, postplanning). The connection between the risks and opportunities is illustrated in Figure 7.2, where the cycle of Figure 7.1, is basically separated into subcomponents of the planned risk (by analysis) and opportunities in the development of risk management. It can thus be said that opportunities are risks managed toward learning and risks are uncertainties directed toward development!

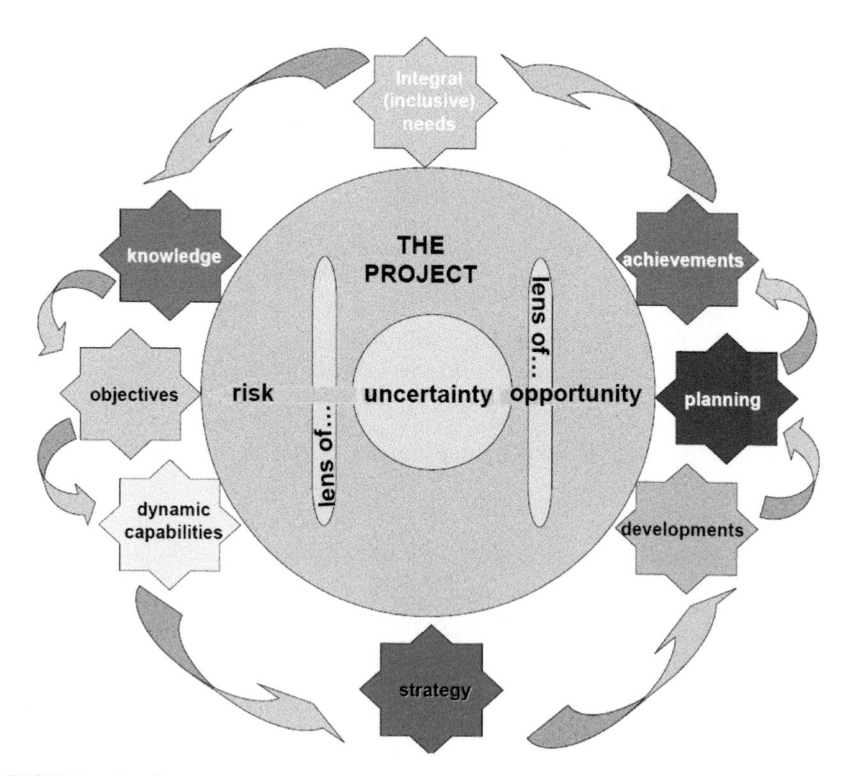

FIGURE 7.2 (See color insert.) The role of risk and opportunity lenses to transform uncertainty into a successful project, following specifically the risk management cycle in Figure 7.1 [inspired and generalized by Chapman and Ward (2011)].

Coming back to *project management* itself (with risk analysis and risk management included), it can be designed onto the two components as the risk and opportunity. According to their degree of accomplishment (low/high) they may lead respectively to a high performance or high knowledge, and are associated to the organization development and learning (Fig. 7.3), as designed in matrices of the "working quadrants" of the project manager (Deming, 1992; Gareis, 2010). Thus, the *checking models, planning models, doing models, and the acting models* can be identified; it is worth mentioning that the difference between action and activity is subtle and it reflects the reality of interaction: "*entering the relationship*" corresponds to action and unfolding the "networking, relationship, connecting, and exploring" in the sense of postmodern management (namely clustering); they are nevertheless different and less effective than "*giving consistency to the relationship*" which, instead, corresponds to acting and "creating value, by exploiting the network and networking" with a focus on extension of acting. The specific result is that *activity is an oriented acting*. From this point, the "circulation" between the values of the management modules of Figure 7.3 provides the characteristics of a successful project: it must be specific, with *measurable* results—thus, comprehensible (tangible or intangible), *acceptable* as costs (so *achievable)*, with *relevance* (economic and social utility, so satisfying sustainability criterion), and finally, being at least durable (so identifiable as achievements in *time*, *time bounding)*. In the ideal case scenario, the regenerative character is the aim: once the achievements reached, they contribute through inclusion to all the successive accomplishments in the specific field of activity. This is what we call a SMART project, here built and guided on the matrix of risks and opportunities, in their turn, as projection of the development and knowledge goals.

Further on, the present essay shows a realistic application of a nanotechnological project, in the domain of graphene-based photovoltaic, in a comprehensible manner *(cognitive analysis),* with identification and management of risks "embedded" in the project management, on objectives and activities (*methodological analysis)*. Overall, we will conclude by going back to the basic ideas related to competitive advantage and its identification through a "creative/interrogative chain" of risks/opportunities analysis. The strategic management establishes, therefore, the optimal strategy (efficient and effective) for developing a sustainable business space (oriented to the final consumer, in a "green" environment).

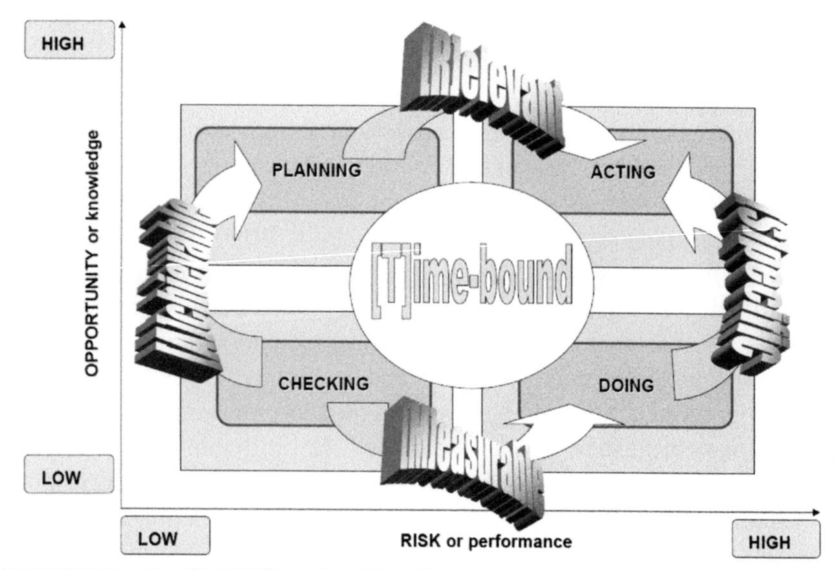

FIGURE 7.3 The [SMART] matrix = "Specific → Measurable → Achievable → Relevant → Time-bound" superposed to [PDCA] cycle = "plan, do, check, and act" (Deming, 1992). Together projected on the grid risk versus opportunity performance versus knowledge, they may ensure a successful project.

Source: Inspired and generalized by Gareis (2010).

7.2 COGNITIVE ANALYSIS

7.2.1 VISION AND MISSION OF A [CASE STUDY] NANOTECHNOLOGICAL PROJECT

As stated by the "Lema of New Technologies" (enunciated by Herbert Kroemer, the Nobelist in Physics in 2000 for pioneering work in the field of Information Technology and Communication) it seems that: *"The main applications of any technology sufficiently new and innovative have always been—and still are—the applications created by that technology."* And yet ... possible current research, innovation, and disruptive technology may occur due to recent discovery of *Graphene,* as a new state of matter (and therefore rewarded by awarding the Nobel Prize in Physics to Andre Geim and Konstantin Novoselov of Manchester University). This is because Graphene was—and still is—a rising "star" on the horizon of material science; with a purely two-dimensional nature (2D), this material has high electronic and crystalline properties, having a resistivity 200 times greater than that of steel, while it effectively conducts *electricity* and heat; Moreover, due to its special electronic spectra the graphene has appeared as the new "relativistic" playground of Condensed Matter Physics, where the physical quantum relativistic phenomena are uniquely combined with new gateways in low-dimensional physics, especially due to its thin dimensional size of a single atom (the carbon). Consequently, while graphene requires no further evidence regarding its importance in terms of fundamental physical phenomenon, "*it is probably the only system in which the ideas of quantum field theory can lead to patentable innovations*"—as claims Frank Wilczek (Nobel Prize in Physics in 2004 for the discovery of asymptotic free bounds on powerful interaction theory). Thus, applications with innovative impact/disruptive using graphene become possible. In particular, there becomes possible to create smart materials with nanoecological, functions, such as photovoltaic systems by sensitization aimed at amplifying the energy conversion of sunlight into green electricity; here, again, its thin structure, made up of a single carbon atom, made the basis of its properties.

Fortunately, this general structure can be combined with the recently advanced notion of the quasi-quantum particle called *Bondon* (Putz, 2010, 2016), in modeling the chemical bond field. It also supports to be generalized under the *Bondots*—eventually emerged from binding of the *coherent* particles of quantum dots (QD) occurring in condensed matter, including on and by the graphenic sheets (Putz et al., 2016). Since QD have already proven useful active centers for solar cells by their sensitization potential, an approach like *double* quantum dots (bondotic; DQD) may further enhance the performances (doubling, at least) of solar photovoltaic energy of the energetic materials working under such a mechanism.

Consequently, the main objectives of such project may unfold as: (1) *Modeling,* (2) *Controlling, and* (3) *Improving (beyond the current performance) for the figures of merit of photovoltaic energy conversion mechanism with DQD (bondotic) activated on graphene; either as functionalized and defective; in subsequent implementation of the transparent electrode, or the active environment in sensitivization of 2D heterostructures for solar cells; accordingly follows the public and intellectual rights objectives under specific* (4) *communication, and* (5) *patenting of the complex system made by combining all of the above goals.*

These objectives are in complete agreement with the current needs of competitiveness of nanoscience and technology. They should contribute to economic demands and social needs of the 21st century, namely: the existence of organic materials (nontoxic, in this case based on carbon) with intelligent mechanisms (here associated with quantum dots), with high renewable energy sources (here, doubling, at least, the performance of photovoltaic energy available today), in a space, as small as possible, (here provided by quantum effects in nanoscale on graphene) and at minimum costs (available as graphene increasingly penetrates the European production circuit).

7.2.2 JUSTIFICATION AND CONTEXT IN A [CASE STUDY] NANOTECHNOLOGY PROJECT

The rationale of this project at European and national levels is summarized in Table 7.1. Other references to sections of the current case study project *"Coherent Nanosystems with Graphene: Towards Double Quantum Dots Sensitive Photovoltaics" here abbreviated as NACOGRAF-2C* applications are given as appropriate.

NACOGRAF-2C project falls within the area of Smart Specialization, falling in selected areas and the business sector of applications, as justified in Table 7.2.

NACOGRAF-2C will contribute, in a first phase, to the development of the western region of Romania by enhancing the research, by generating the present solution, that is, the double points-graphene-photovoltaic system (at reaching the objective) increasing to at least doubling the conversion of light into electricity. As a first impact on the following identified strategic regional objectives (as monitored by the regional Development Agency—western region of Romania) are, for example (see also Table 7.3):

- increase the capacity for research and innovation in the western region of Romania (through the DQD in graphene with applications on solar cells)
- develop existing infrastructure and centers of excellence (to be created and developed through this project)
- support relationships between research and development institutes—business (through further implementation of this nanoquantum-PV process in obtaining photovoltaic cells as viable product, with economic impact)

TABLE 7.1 The Strategic Context of a Project (Case Study on this Project NACOGRAF-2C) of Nanotechnology Research and Innovation in the Framework of National Competitiveness, Included in the Action Oriented to Knowledge-based Economy, as in the Horizon 2020 Program of the European Union, with Impact Far Beyond it.

Context	Crit.	Strategic description	The present (NACOGRAF-2C) project rationale
European	EU_1	The strategy "Europe 2020" established strategic directions, among which "*the intelligent growth based on knowledge and innovation*" for preventing the continental structural differences—especially avoiding a further economical global crisis consequences in Europe	NACOGRAF-2C advances a *disruptive* nanoscience on amplifying the solar cells efficiencies based on quantum field theory (double quantum dots effects in special) applications on graphene
	EU_2	The 2011 European Union initiative "*An Innovation Union*" targets the growth of the research and innovation strategies as an ex-ante conditionality for whatever access to the structural funds for 2014–2020 period	NACOGRAF-2C project is *intrinsically* structured on novel fundamental research and experimental development of innovative process, that is, the double quantum dots sensitization of the photovoltaics working with graphene
	EU_3	The coordination and *synergy* in the framework program Horizon 2020 are understood and specified in the Annex 1 of Reg. (CE) no. 1303/2013, so defining the Cohesion Politics of European Commission—COM (2011)615 for 2014–2020	NACOGRAF-2C addresses a thematic research of broad interest at local, regional and international level, so *synergistically* approaching social and economic consequences of project results regarding the eco-nanoimplementation of coherently quantum-based solar cells in privet and public sectors
	EU_4	The "Europe 2020 Competitiveness Report: Building a more competitive Europe" (2012) specified that, for Romania, about 70% of the total employers from industry (about 1.2 million) are in *less competitive* sectors	NACOGRAF-2C results in coherent activation of photovoltaics may have (upon its unfolding activities, i.e. in its post-project phase) the same effect and impact as the control of coherent light sources led to the LASER discovery. The *follow-up disruptive applications* in all economic and social levels should increase sustainability of input energy on the medium- and long-term research and development impact in industry

TABLE 7.1 *(Continued)*

Context	Crit.	Strategic description	The present (NACOGRAF-2C) project rationale
	EU_5	The "Innovation Union Scoreboard" of European Union (2014) includes Romania in the category of *modest innovators*, the last of the four categories considered	NACOGRAF-2C may consistently contribute, by running the excellence research project with the top equipment at the R&D Institute which implements it (see also, Chapter 11 of the present monograph). The dynamic team is required to be conducted by indeed high-level capabilities. From ex-post project perspective, further impact of economic competitiveness and social welfare is envisaged
Romanian	RO_1	The National Strategy of Research, Development and Innovation 2014–2020 and the National Strategy of Competitiveness have on the forefront thematic objectives the "*research consolidation, technological development, and innovation.*"	NACOGRAF-2C jointly fulfill these needs by *innovative* (fundamental and experimental) *research* addressed, as well as *consolidating it by continuing* at Research, Development, and Innovation center/institute another project with Glocal, regional relevance, for example, "Laboratory of renewable energies—photovoltaics" (see also, the Chapter 11 of the present monograph)
	RO_2	The (National) 2014–2020 Strategy of Research, Development and Innovation and the (National) Strategy of Competitiveness have on the forefront priority axis of investments "*the consolidation of research and innovations, of infrastructure and of capacity in developing excellence in the field of Research and development, as well as the promotion of excellence centers especially those of European interest*"	NACOGRAF-2C fully responds to these purposes by addressing excellence in nanoscience, with top equipment at implementing Research, Development, and Innovation center/institute, with top expert in nanoscience from (European) academic and industrial space leading the implementation of the research project and a specialized team oriented to this aim

TABLE 7.1 *(Continued)*

Context	Crit.	Strategic description	The present (NACOGRAF-2C) project rationale
	RO_3	Competitiveness Operational Program (COP) 2014–2020 with its Action "attracting high-level personnel from abroad in order to enhance the RD capacity", especially, envisages "*the increase of the (Romanian) research at the European Union level*"	NACOGRAF-2C developing research activities and, especially, the experimental developments will lead the whole project team to concentrate the excellence in the Research, Development, and Innovation center/institute, regarding the photovoltaic studies with graphene while adding the needed pressure for valorizing the R&D results in high-level publications, patens, and further research proposal to Horizon 2020
	RO_4	According to Competitiveness Operative Program, with the first version for consulting, elaborated by Ministry of European Funds, Romania has the lowest percent of *R&D personnel* from the total employers (0.46% in 2011 by National Institute of Statistics), being nevertheless *decreased* from 10,300 in 2005 to 5000 in 2012 (equivalent full norms) and with a regional (except Bucuresti–Ilfov region) distribution below the average level of Europe 27	NACOGRAF-2C will *increase* the National Institute of Research— Development for (NIRD) mass of *highly qualified (PostDoc) researches* with full norms, by *two units in the project*, and with perspective of further increasing post-project: by the novelty, by excellence, and increasing the importance of the research developed by appropriately communicated and disseminating though proximity industry, by awareness on the academic media (universities and other research and developments units in the western region of Romania, and at national level, Bucuresti–Ilfov included)
	RO_5	The Romanian Law no. 83/2014 considerably improved on the *intellectual properties* so clarifying the legal rights in valorizing (also financially) the rights of the employers brevetting in the name of institutions	NACOGRAF-2C will patent at least two national (+ one European patent submission) experimentally developed processed in sensitizing-to-amplifying the photovoltaic performances by coherent double-dots mechanism on graphene, so highly stimulating its personnel contributors toward *valorizing further rights* in relation with direct and indirect beneficiaries, either economic or industrial, in Romania or Europe

TABLE 7.2 Scientific-Technical and Socioeconomic Implications of the Present Case Study Nanotechnology Project (NACOGRAF-2C).

Area and subarea of the project	Eco-nanotechnologies and advanced materials	NACOGRAF-2C project uses the coherent quantum dots and their bindings on graphene through the bondots, so inherently addressing a process on a nanoscale (by quantum) and of essentially nontoxic/eco-nature (by carbon-based graphene)
	Materials	NACOGRAF-2C uses graphene as the main component of envisaged 2D stacking heterostructure, which is worldwide called as a "miraculous new material in 21st century!"
	Materials for energy	NACOGRAF-2C research output is converting the solar energy into electric energy at high rate of transformation, so fully approaching an energy renewable process
Economic sector	Innovation, technological development, and plus value	NACOGRAF-2C has as the direct applicability the innovative technology of using double quantum dots (in quantum coherence) enhancing-to-amplifying the producing electricity by solar light so adding a plus value to the available solar cells and photodetectors on the market
	Energy and environment management	NACOGRAF-2C envisaged objective and results in highlighting the electric energy by quantum coherence on embedded graphene in photovoltaics fully conforms with the national strategy of competitiveness in the field of renewable energy and lesser energy from fossil or other environment sources (gas, water, and wind) so lesser-to-zero affecting the environment by its further applications

TABLE 7.3 Relationship of NACOGRAF-2C with Other Programs/Strategies/Projects/Other Relevant Documents.

Crt. no.	Type (program/ strategy/project/others)	Name	Relation
1.	Regional Operational Program (ROP)	The Strategy for Regional Development of West Region in 2014–2020	Services for increasing of competitiveness and intelligent specialization in the western region

- the internationalization of research, development, and innovation (specific to this project, with great perspectives for extension in post-project stage, with the subsequent increase of the research team and RandD infrastructure)

- encourage intellectual property (in this project will be designed at least two national patents and a European patent application)
- encourage technology transfer (based on skills acquired by studying and experiencing ongoing process graphene–quantum photovoltaic system and designing a viable product (solar cell) in post-project) and so on.

7.2.3 MANAGEMENT OF A [CASE STUDY] NANOTECHNOLOGY PROJECT

The success of a nanotechnology project in general and the present (NACOGRAF-2C) in particular is provided by systematic management monitoring, control, and evaluation of results; all will be judged by the successful application of the following criteria:

- compliance with *time* schedules for implementation of each activity and subtasks;
- compliance with the *budget* for each activity, with implementation of specific subtasks;
- efficient exploitation of *resources,* both of the human/staff and equipment available, logistics and locations of the research, development, and innovation institute;
- regional/national/international *perception* on the development of the project.

The project is closely supervised by the management team in cooperation with the Project Director at all stages of implementation of specific activities and sub-activities, minimizing risk occurrence. The functions of the team manager are summarized in Figure 7.4 which comprises general routes to maintain high standards of the project by achieving ambitious targets and the performance of the settled indicators, synergistically as: (1) within the planned time, (2) within the available budget, (3) with highly skilled implementation team, and (4) ensuring client satisfaction by providing competitive products and post-project relations.

However, the management team together with the Project Director will face all challenges occurred during the development (Fig. 7.5) and *implementation*, for example:

- create the link between project team members and other members who may have different perspectives on the scientific nature of the study concerned, cooperation with new people (on staff to be employed on new positions created in the project)
- maintain relationships and creating an institutional culture both internally and in relation to foreign partners (regional, national, and international)
- ensure good connections with the funding bodies—Ministry of Education and Scientific Research—National Authority for Scientific Research and Innovation, certification bodies at home and abroad;
- maintain proper presentation of the status and objectives of the project in the media and for the target public
- perform public acquisitions in compliance with the law;
- solve ad hoc circumstances (which cannot be predicted) of social/human, technical, economic type, and even direct or indirect political influence on the smooth running of the project activities.

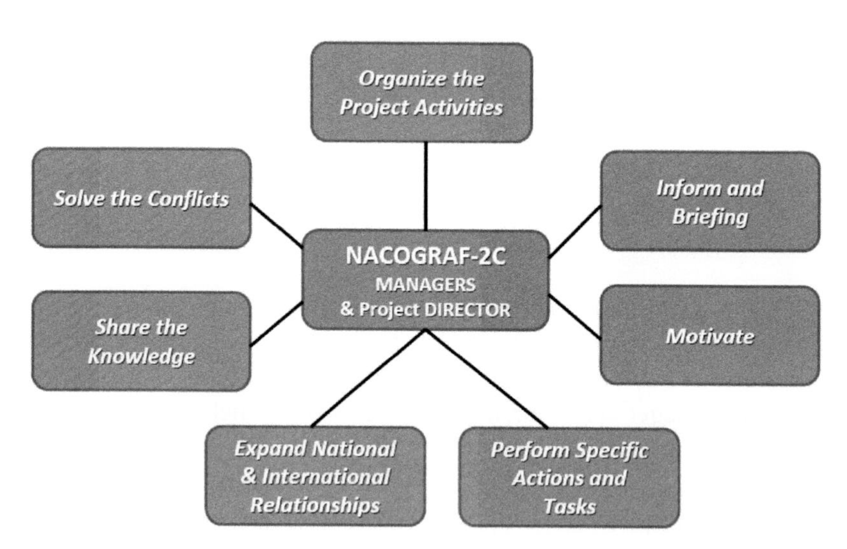

FIGURE 7.4 Functions of managers and project director in NACOGRAF-2C.

Regarding the *monitoring* of project activities, for the successful development of sub-specific activities and the overall project, there will be combined the following management techniques:

- *periodic meetings*, ensuring that each sub-activities and tasks to be performed are well anticipated and prepared;
- *pareto charts* to isolate 20% of factors/parameters responsible for about 80% of planned actions—and address them first;
- *flowcharts* for measuring the *effectiveness* of actions taken (by reporting the effects recorded compared to the effort implemented) and *efficiency* by comparing actual results with planned ones.

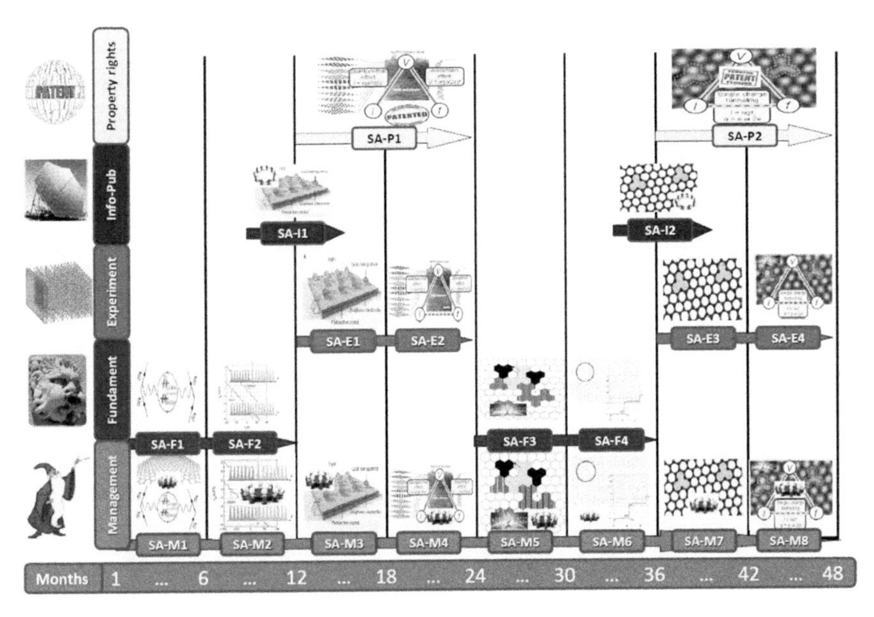

FIGURE 7.5 (See color insert.) The structure and development of activities and sub-activities ("SA") breakdown structure and flow of NACOGRAF-2C project.

Equally important is the control of the project implementation, which will be done as follows:

- *periodic inspections* of staff and equipment involved (spot checks)
- *control charts* to graphically highlight the errors due to documentation, calculations, measurements, while applying the techniques of comparative analysis and cross-checking, adjusting them by feedback control and repetition of specific operations and so on

- *trend analysis* to identify the mathematical on mathematical basis the qualitative evolution of tasks and subtasks of the planned activities and sub-activities
- prepare *interim reports/internal* on the economy, participation, of acquired knowledge, and sufficient accomplishment of activities through results obtained.

Overall, all the key aspects needed to fulfill the project objectives are contained in this integrated management which covers all nine basic components (in compliance with the model provided by the Institute of Project Management, USA) necessary for a research and development successful project, namely: (1) harmonizing planned activities, (2) covering the entire provisioned area of research, (3) integrating subtasks within the time set, (4) the rationalization of cost, (5) maintaining the quality of the results, (6) classification of human resources, (7) carrying out safe procurement, (8) ensure optimum communication, and (9) minimizing risks (see the next section).

7.3 METHODOLOGICAL ANALYSIS

- In any human activity, risks are inevitable so that development and implementation of project NACOGRAF-2C cannot be risk-free. However, in contrast to uncertainties (which cannot be controlled), risks can be identified a priori and a scenario can be made to counter them. For this project, we proceed as follows:
- identify *types of risk:* technical (T), social/human resources (S or H), economic (E), and political (P)
- quantify the *possible sizes of risk*: marginal/safely (pt. = 1) minor (pt. = 2), moderate (pt. = 3) large (pt. = 4), and catastrophic (pt. = 5)
- determine *risk management*: risk avoidance, reduce sensitivity to risk, risk transfer, and counter/acceptance of risk.
- Based on these considerations, further analysis of risk analysis will be done and we will estimate the degree of risk, both for each work task associated to each activity (refer to Fig. 7.5), and for the whole project, Table 7.4 and Figure 7.6.

Naturally, for risk management to be effective, its perception must be dynamic, that is, depending on the time schedule of the project.

TABLE 7.4 Risk Analysis for Project Activities and Sub-activities NACOGRAF-2C of Figure 7.5.

Action	Sub-action	Risk identification	Risk impact	Pts.	Risk approach	Type
Fundamental research	SA-F1	Approximations in theoretical modeling and limits of computing	Marginal	1	*Avoided* by cross-checking or calibrating in between analytical and numerical analyses	T
	SA-F2	Limiting computing capability	Marginal	1	*Transferred* risk by connection with computing cluster available in the NIRD-Timisoara and, at the limit, to the publicly available Blue-Gene facility in West University of Timisoara (as one of the potential beneficiary of the project)	T
	SA-F3	The need for combined computation software	Marginal	1	*Avoided* by acquisition of the required computer programs	T
	SA-F4	The running time for simulating many photovoltaic systems may be longer than expected	Marginal	1	*Transferred* risk by connection with cluster computing at NIRD-Timisoara and, at the limit, to Blue-Gene facility in West University of Timisoara (as one of the potential beneficiaries of the project)	T
Experimental development	SA-E1	Synthesis of functionalized graphene and its embedding into stacking stable 2D heterosystem	Moderate	3	*Reduction of the sensibility* to the risk by (pre)acquiring appropriate scientific benchmarking, trainings and services	T
	SA-E2	Measuring with metrological accuracy the photovoltaic effects on functionalized graphene and observing the coherent double quantum dots effects (bondots) reality	Minor	2	*Avoided* by combining the dedicated advanced equipment in NIRD-Timisoara/ Laboratory of Renewable Energies-Photovoltaic to synergistically work on the metrological measurements needed	T

TABLE 7.4 *(Continued)*

Action	Sub-action	Risk identification	Risk impact	Pts.	Risk approach	Type
	SA-E3	Producing the stable atomic/bonding vacancies and associated molecular holes in nanoribbons sheets, while considering it either as electrode or as active medium into composed heteromaterials manifesting intra-van der Waals interactions among 2D nano-components	Moderate	3	*Reduction of the sensibility* to the risk by (pre)acquiring appropriate scientific benchmarking, trainings, and services	T
	SA-E4	Measuring with metrological precision the photovoltaic effects on defective graphene while observing the double quantum dots effects (bondots) reality	Minor	2	*Avoided* by combining the NIRD-Timisoara/Laboratory of Renewable Energies-Photovoltaic nano-tools to synergistically work on the metrological measurements	T
Property rights	SA-P1	The prolongation of the bureaucratic formalities in patenting registration, publication to its awards at State Office of Inventions and Trademarks (O.S.I.M)	Marginal	1	*Avoided* by covering the entire fee taxes from the initiation of the brevetting request from NIRD-Timisoara side by a rapid response (respecting all legal terms) of any notifications sent by OSIM	T
	SA-P2	Legal issues with national (Romanian or Italian) versus European rights covering the defective graphene as solar cells sensitizer with double quantum dots at European Office of Patenting	Marginal	1	*Transferred* to European level by signing additional needed forms in between the NIRD-Timisoara and an international Chemical Research center (from where the project Director originates)	P

TABLE 7.4 (Continued)

Action	Sub-action	Risk identification	Risk impact	Pts.	Risk approach	Type
Information and publicity. public Aquisitions	SA-I1	Prolongation in time of acquisitions and publicity by changing in the law of public acquisitions	Marginal	1	*Transferred* to the immediately available next sub-activity, that is, by unifying SA-I1 with SA-I2 sub-activities	E
	SA-I2	Exceeding the to-date available budget	Marginal	1	*Reduction of the sensibility* to the risk by reducing the expenses and covering the exceeding budget by NIRD`s own funds	E
Management of the project	SA-M1	The possibility for non-presenting of the virtual candidates for occupying the available four-vacant PostDoc positions.	Marginal	1	*Reduction of the sensibility* to the risk by the scientific manager of the project at NIRD-Timisoara to the actual PhD students which will be oriented for Post-Doc continuation on NACOGRAF-2C project	H
	SA-M2	Possible season (celebrations, strikes, etc.)-related limitation to travel for benchmarking training and scientific communications	Marginal	1	*Transferred* to the immediately available next sub-activity, that is, by unifying limitative actions in SA-M2 to the SA-M3 sub-activities	S
	SA-M3	Possible limitative funding in this phase by the available transferred project budget to-date	Marginal	1	*Transferred* to the NIRD-Timisoara temporary assistance with the required budged by its other funds, possibly bank loans, for normal flowing of planned actions	E

TABLE 7.4 *(Continued)*

Action	Sub-action	Risk identification	Risk impact	Pts.	Risk approach	Type
	SA-M4	Possible limitative personnel fully engaged in this phase of the project when continuous metrological measurements are necessary	Minor	2	*Avoided* by earlier planning of all project members activities in the anticipated period of measurements so that all personnel not be involved in other research–development activities in various projects they may be co-involved	H
	SA-M5	Possible limitations in available commercial algorithms due to the high specificity of the defective graphenic structure simulation	Marginal	1	*Avoided* by self-composing the computational algorithms with the specialized personnel from the project team (the leader and his PostDocs)	T
	SA-M6	Possible to many invitations in attending public presentations and conferences will impede the focus on the second part of the project	Marginal	1	Avoided by extending the invitations to the members managers of the project not directly involved in the actions of this sub-activity	S
	SA-M7	Possible new regulations about the electron beam intensities and safety	Minor	2	*Reduction of the sensibility* to the risk by changing the investigation technique by pulses and/or by continuous irradiation with lower intensity yet not affecting the stability of the investigated sample	P
	SA-M8	Possible new formal/institutional conditions in applying the Horizon 2020 with the post-project development	Marginal	1	Transferred to the NIRD-Timisoara management in finding the appropriate program, including by signing new cooperation agreements with strategic beneficiary of the present project (national, regional, or European, as deemed), engaging new institutional investments in this regards, and so on	P

Accordingly, Figure 7.6 includes the data in Table 7.4 as histograms. From here, one can evaluate the median of global risk for the project; Thus, it can be considered that the action Property Rights (SA-P1 and SA-P2) occurs in two management activities (two-time modules, of 6 months each) being summed as twice the median risk for the project NACOGRAF-2C. For example, the amount of points considers two additional points in the total of risk points, being reported to 22 entries.

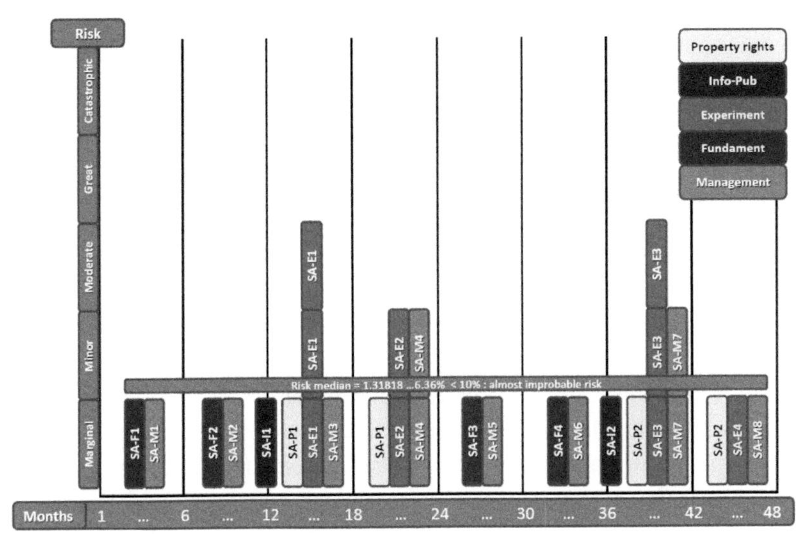

FIGURE 7.6 (See color insert.) The graph of risk dynamics for project NACOGRAF-2C as histogram based on the risk analysis carried out in Table 7.4.

Thus, the median time to risk for NACOGRAF-2C was determined to be 1.31818; following this assessment, for the project in general, with all its actions and sub-actions, risk is ranged from marginal and minor risk, in the proximity of the marginal limit. Now, considering the extent of the risk occurring in cases 1≤ risk point ≤5, the current percentage (probability) will correspond to a probability of occurrence of a minor risk of 6.36% (Fig. 7.6); furthermore, this estimation it is placed in category *almost impossible to see a risk* (as classified in the literature, any risk of a percentage ≤10%). In addition, since there have been taken into account all the risks and countermeasures have been taken by appropriately planned measures, we can say that in terms of management and risk analysis this project is very safe, and in socioeconomic terms—viable and sustainable.

7.4 CONCLUSIONS AND PERSPECTIVES

Back to the fundamental issue of strategic management: how do we achieve, develop, and preserve competitive advantage—here through the lens of risks and opportunities? The author has formulated and is in the process of developing this response, by using the conceptual approach through the business development model with the distinctive advantage cube strategy (Putz, 2017, 2019a,b), and applied it to nanotechnology clusters (Putz, 2019; Putz and Petrişor, 2020). The cube of competitive, sustainable, and regenerative advantage combines these three fundamental features of a successful business by generating a cube whose nodes correspond to eight types of business models, multiplier and with the ability to cover the whole business space, in the econo-sphere. Similarly, in the case of nanotechnology projects, the business is the project itself and the characteristics of competitiveness, sustainability, and regenerative advantages have their correspondence in the basic principles of eco-ecology, namely:

- The "*polluter pays*" [invests] principle "on the account of" competitiveness, as it should meet international requirements, standards, and environment protection requirements; In accordance with this principle, physical and chemical experiments conducted in compliance with the regulations on wastewater discharge, vapor release in the atmosphere, and so on; they all are consistent with the preservation of heritage and natural resources, starting with the protection of the health of researchers or citizens directly or indirectly affected by the project or post-project developments (tacitly respected by the NACOGRAF-2C project).
- *Sustainable development,* fully respected by the NACOGRAF-2C project, is in line with the National Strategy for Sustainable Development of Romania Horizon 2013–2020–2030 because it acts by directly and indirectly including *horizontal measures for the principle of development*: (1) The project is fully oriented toward *improvement of renewable energy* (through solar-to-solar electricity solution for photovoltaic systems that work with DQD consistent with graphene as an electrode or photoactive media); (2) *The project maximally avoids the destruction* or alteration of the environment (as the main source of energy is sunlight, without involving, for example, water, waste gases, or wind); (3) The project produces results that will greatly improve *the efficiency of*

buildings, public lighting of strategic institutions (such as hospitals, schools, other research activities, economic and administrative spaces), and street lighting with utmost effectiveness; (4) The project advances an essential *nontoxic solution* for *solar cells* by using graphene—made exclusively of carbon, especially through the results of the second part of the project, and in the post-project phase in which only defective graphene is involved; (5) The project *extends the life cycle* of solar cells and sustainability through the exclusive involvement of coherent quantum effects in intelligent materials (functionalized and defective graphite embedded in 2D controlled heterojunctions), compared to current electrodes based on organic–inorganic compounds; (6) The project *informs the media and the public* through information and publicity activities on the quantum (nanointelligent) solution, safe for photovoltaic materials and for the next generation of solar cells resulted from them, thus contributing to the creation of a "consciousness" of smart public consumption, the green labeling, and awareness of direct and indirect beneficiaries, businesses, and investments (including green procurement) implementing green energy science and technology.

- *Equal opportunities* ensure the continuity of the nanotechnology project. It is thus being disseminated and taken over, and then integrated into subsequent nanotechnology projects and developments, so ensuring: (1) transparency and cross-disciplinarity; (2) *inter- and intra-generationality equilibrium:* especially through young specialists in certain areas, even though many of them are in the early stages or scientific research career advancement; (3) the *nondiscrimination of gender, race, religion (including atheism), disability, age, sexual orientation, ethnicity, nationality, condition, or social origin;* (4) *the time flexibility and project-specific program* so that the staff and resources involved can be used to their full potential in multi-project activities, multi-portfolio projects, and multi-stakeholder and businesses activities; (5) *the meritocratic selection* of the workforce involved and the appropriate reward, including salary, for achieving objectives and overcoming the problems; and (6) *objectification and optimization* of public procurement actions, in all phases of implementation, post-projecting, with as much possible correlation to open access to information/solution or facilitating joint venture for further developments.

Finally, we can construct the cube of the distinctive advantage based on the triplet ("pollution/investment, sustainability, and equality of opportunity"). As the cube of competitive, sustainable, and regenerative advantage generates eight models of 3D business (Putz, 2019a,b); so the current cube of research–development–innovation generates, from the perspective of risk exploration and exploitation of opportunities, the conversion of uncertainty into efficiency and effectiveness. This distinctive cube is, therefore, a cube of risk and opportunity identification and management, and the space covered thereby matches the answer to eight fundamental, generalized questions (Chapman and Ward, 2011):

1) **When?**: Strategic question (time related), specific to business/business/wise projects (win–win–win of the basic triplets above, or, in business related to the triple contractor–contractor—environment)
2) **In what way (*which way*)?**: technical question, specific to relationship and network planning, contracting, case studies, operation (*loss–win–win*)
3) **Who?**: a resource question, clarifying contributing parties, ensuring success in the sense of project/business profitability *(win–lose–win)*
4) **Who for/whose?**: a utilitarian question, specific to sustainability, econo-ecology in general and smart business/projects in particular *(win–win–lose)*
5) **What?**: is the question of modeling, design, documentation, but also with post-project exploitation potential and deviation in negative impact if it is not controlled or relaunched in the post-project (*lose–lose–win*);
6) **Where?**: Exploratory question, reveals availability of outsourcing of actions, contextualization of objectives and operations, corresponds to the *outsourcing strategy,* and also opens toward *open access or open data,* even with high environmental costs (*lose–win–lose*)
7) **What with?**: a question specific to polemocratic affairs (*win–lose–lose games),* identifying the key resources to ensure the planned gain, an opportunity for overdevelopment and "overheating" of resources
8) **Why?**: the regenerative motivational question ensures the continuous knowledge, but also the impulse to unconditional success, and often to step to failure by playing with a negative sum, the anteroom of the "animal spirits" type business/project "animal spirits" (*lose–lose–lose).*

Going through this 3D chain of questions, with appropriate responses as strategic games, by respective color code shown in Figure 7.7, see also Figure 7.2 of the Introduction, one may generate a "*specific Hamiltonian path*." Along such a path, each state is reached once only, while a complete life cycle of the project is ensured. Moreover, basically, within such framework, all the risks and opportunities were identified, the potential has been activated, and the reinvention on the second curve/development cycle or in a new project is prepared by multiplying the added value. The plus-knowledge, between "good and evil" (+ and − of strategic cube), between probable and unlikely, between accomplished and unaccomplished goals respecting the wise (ideal) business, between synergism (creative union) and stigmergism (the creative union with memory)—they are working in a meta-equilibrium and a meta-networking (nanotechnology) dynamics with the triple-fold econo-ecological feature: competitiveness, sustainability, and regenerative advantages (Putz, 2019; Putz and Petrişor, 2020).

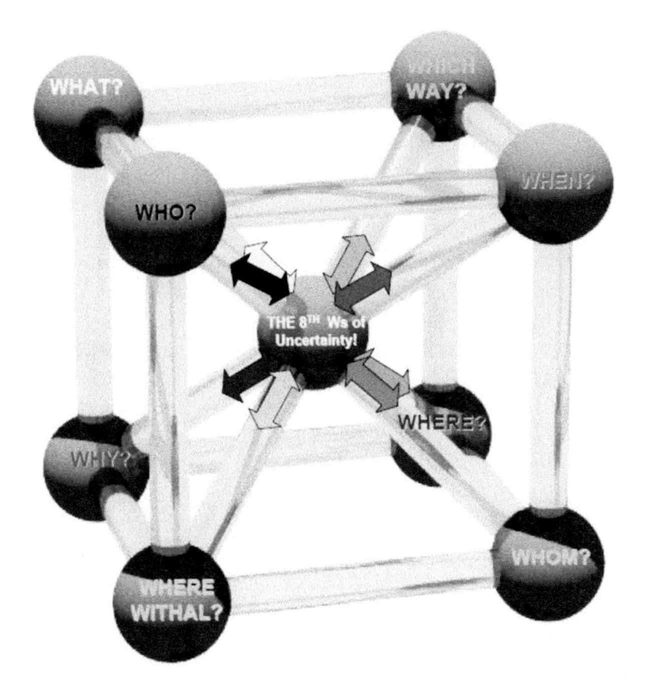

FIGURE 7.7 The strategic cube of competitive advantage complemented with the Ws (eight fundamental questions) of risk management, for successful projects.

Source: Inspired and generalized according to Chapman and Ward (2011).

KEYWORDS

- **risk analysis**
- **risk management**
- **project management**
- **nanotechnological project**
- **competitive advantage**

REFERENCES

Chapman, C.; Ward, S. *How to Manage Project Opportunity and Risk—Why Uncertainty Management Can Be a Much Better Approach Than Risk Management* (The updated 3rd Edition of Project Risk Management); Wiley & Sons: Chichester, UK, 2011.

Deming, W. E. *(Crisis) Out of Crisis*; Cambridge University Press: Cambridge, 1992.

Gareis, R. *Happy Projects. Managementul proiectelor şi programelor. Managementul portofoliilor de proiecte. Managementul Organizaţiilor orientate pe proiecte [în original Happy Projects!, 3. Auflage, Manz'sche Verlags und Universitätsbuchhaltung GmbH, Wiene, 2004]*; ASE Publishing House: Bucureşti, 2010.

Hristova, I.; Ilić, I.; Kocjančič, A.; Struna, D.; Zajc, M. D. *Beyond the Horizon. Practical Guide to Developing Competitive Project Proposals in Horizon 2020*; RR&Co. Knowledge Centre Ltd.: Ljubljana, Slovenia, 2014–2015.

Porter, M. *On Competition*; Harvard Business Review Press: Brighton, 2008.

Putz, M. V. The Bondons: The Quantum Particles of the Chemical Bond. *Int. J. Mol. Sci.* **2010,** *11* (11), 4227–4256; DOI: 10.3390/ijms11114227 (Special Issue: Atoms in Molecules and in Nanostructures; M.V. Putz, Guest Editor); WEB: http://www.mdpi.com/1422-0067/11/11/4227.

Putz, M. V. *Quantum Nanochemistry. A Fully Integrated Approach (5 Volumes Package): Vol. I. Quantum Theory and Observability; Vol. II. Quantum Atoms and Periodicity; Vol. III. Quantum Molecules and Reactivity; Vol. IV. Quantum Solids and Orderability; Vol. V. Quantum Structure-Activity Relationship (Qu-SAR)*; Apple Academic Press & CRC Press: Toronto, Canada–New Jersey, USA, 2016; pp. 3086+index; ISBN: 978-1-771881-38-8, http://www.appleacademicpress.com/title.php?id=9781771881388.

Putz, M. V.; Tudoran, M. A.; Mirica, M. C. Quantum Dots Searching for Bondots: Toward Sustainable Sensitized Solar Cells. In *Sustainable Nanosystems Development, Properties, and Applications*; Putz, M. V.; Mirica, M. C., Eds.; Editori: IGI Global: Pasadena, California, USA, 2016, DOI: 10.4018/978-1-5225-0492-4, Ch.9, http://www.igi-global.com/chapter/quantum-dots-searching-for-bondots/162091.

Putz, M. V. Strategic Cube of the Organization Competitive Advantage. [Originally in Romanian: Cubul strategic al avantajului competitiv al organizaţiilor]. MBA Thesis, Faculty of Economy Science and Business Administration, West University of Timişoara, 2017.

Putz, M. V. *Scientific Entrepreneurship by the Strategic Double Cube of Competitiveness–Knowledge Transfer*, Chapter 5 of the Present Monograph, 2019a.

Putz, M. V. *The Strategic Cube of the Distinctive Advantage. Epistemological Approach*, Chapter 2 of the Present Monograph, 2019b.

Putz, M. V. The Strategically Dynamics of the Research–Development–Innovation Potential in the Nano-Technological (Meta)Clusters [Originally in Romanian as: Dinamica strategica a potentialului de cercetare-dezvoltare-inovare in (meta)clustere nanotehnologice.] PhD Thesis, Faculty of Economics and Business Administration of West University of Timisoara, in Preparation, 2019.

Putz, M. V.; Petrişor, I. *The Code of Strategic Management: From Modern to Post Modern (Meta) Clustering Approach*—in Preparation, 2020.

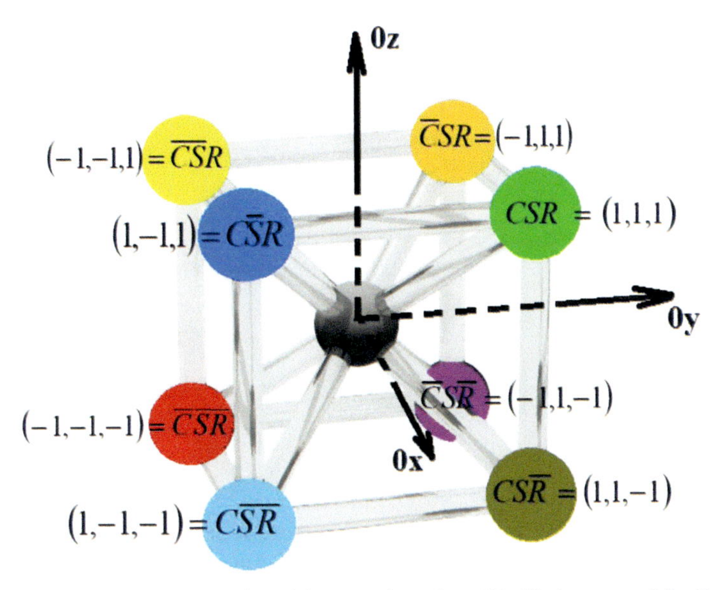

FIGURE 2.2 The strategic cube with strategic points (1)–(8) interpreted in Cartesian coordinates 3D in relation to the central point of coordinates (0,0,0).

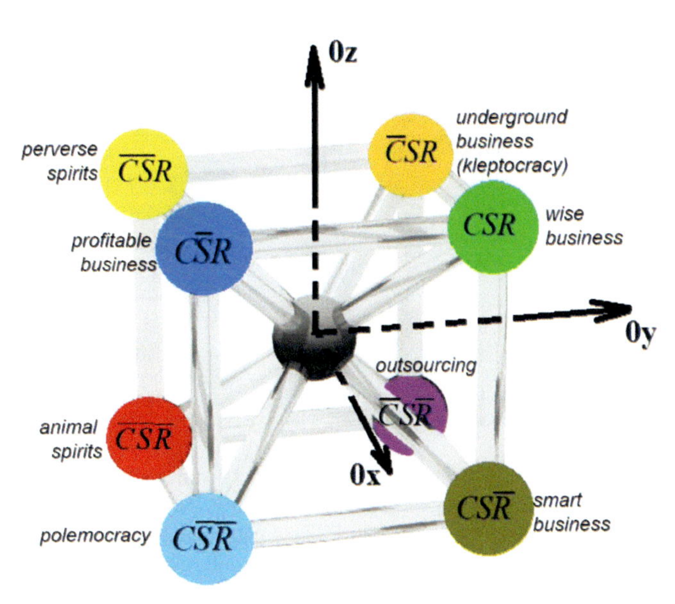

FIGURE 2.3 The strategic cube with the strategic attributes corresponding to strategic points in Figure 2.2.

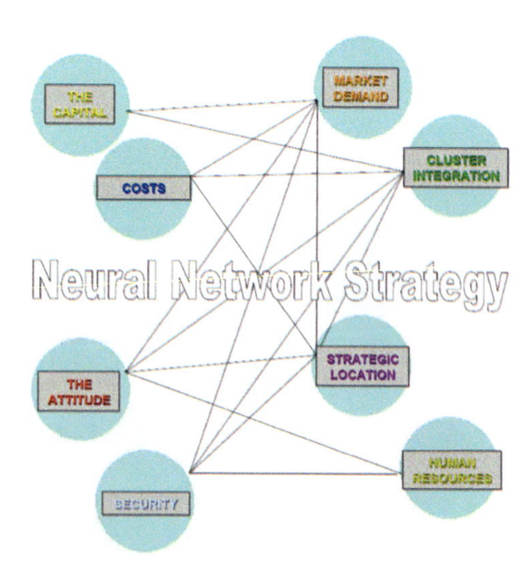

FIGURE 6.3 Strategic management of networking as a linkage (not necessarily with networking consistency) in the multinodal logic (logistics); note the catalyst factors (as integration in conglomerates–clusters, the market demand, the human resources, and the strategic placement) along the constraint factors (as the cost, the capital security, and the strategic attitude) in the space of business development; here generalized according to Porter (1995).

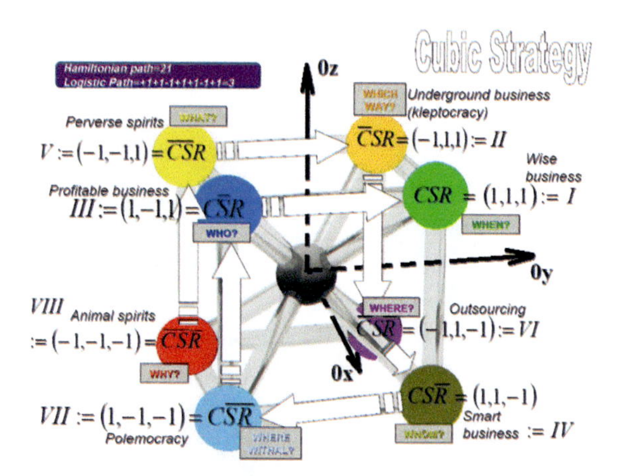

FIGURE 6.4 The 3D strategic management of the distinctive advantage cube of the competitive, sustainable, and regenerative advantage, as developed with the multinodal course of dual quantification: by (1) the longest path, being self-referential, along the so-called Hamiltonian path; and by (2) the logistic path, exo-referential, and positively oriented to satisfying/reaching the competitive (on 0X), sustainable (on 0Y), and regenerative (on 0Z) advantage. It nevertheless gives consistency (or effective relationship) to 2D connectivity of the network of the flat strategy of Figure 6.3.

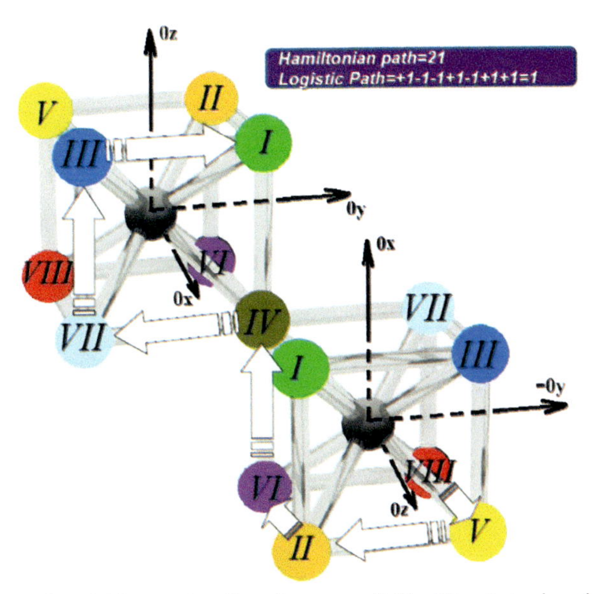

FIGURE 6.5 Uni-nodal interaction (by a "common tip") of the strategic cubes in order to achieve a competitive, sustainable, and regenerative business, with the same Hamiltonian path as in the endogenous cube of Figure 6.4, but with an optimized logistic path.

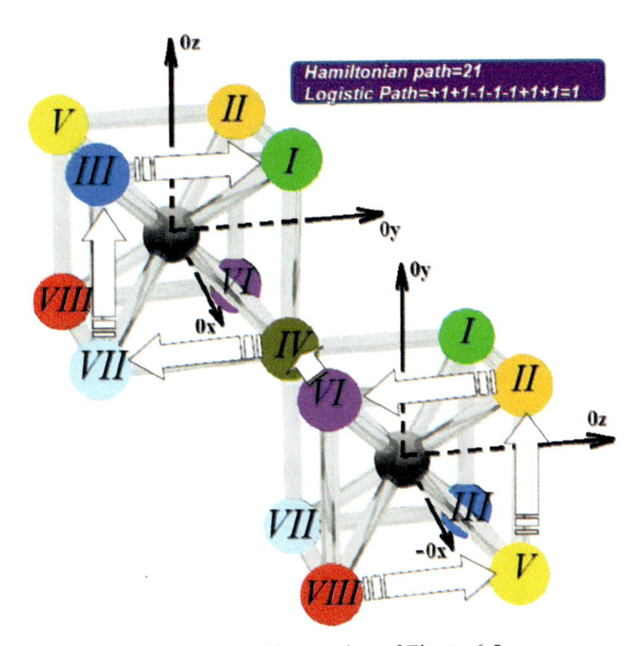

FIGURE 6.6 Variant of the uni-nodal interaction of Figure 6.5.

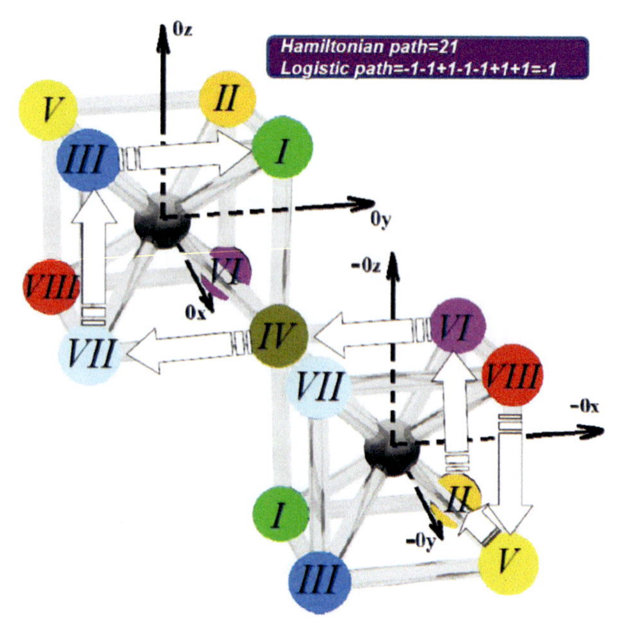

FIGURE 6.7 Variant of the uni-nodal interaction of Figure 6.5, but with a minimized logistic path.

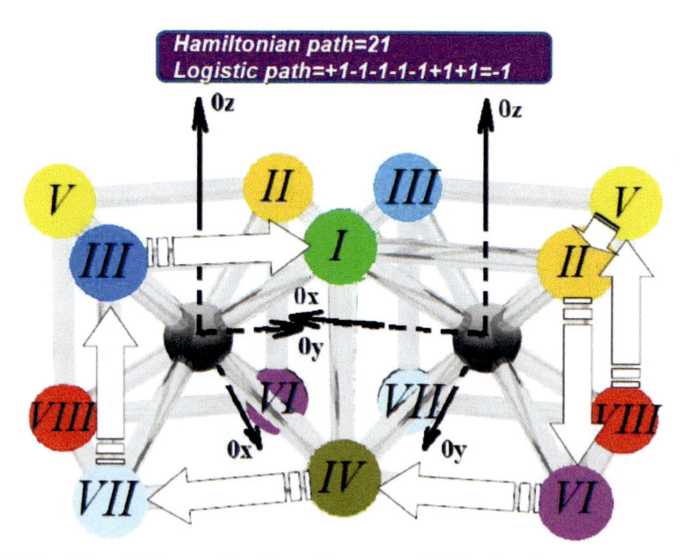

FIGURE 6.8 Bi-nodal interaction (by a "common side") of endogenous strategic cubes of Figure 6.4, with the same Hamiltonian path, but with a minimized logistic path—at the value in Figure 6.7.

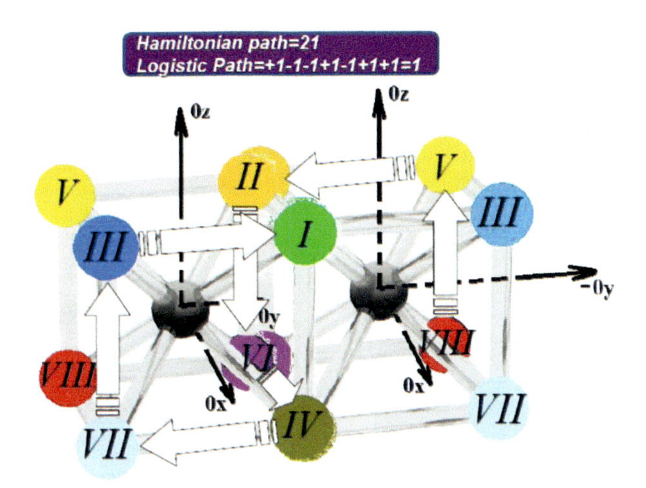

FIGURE 6.9 Bi-nodal interaction (by a "common side") of endogenous strategic cubes of Figure 6.4, with the same Hamiltonian path, but with an optimized logistic path—at the value in Figures 6.5 and 6.6.

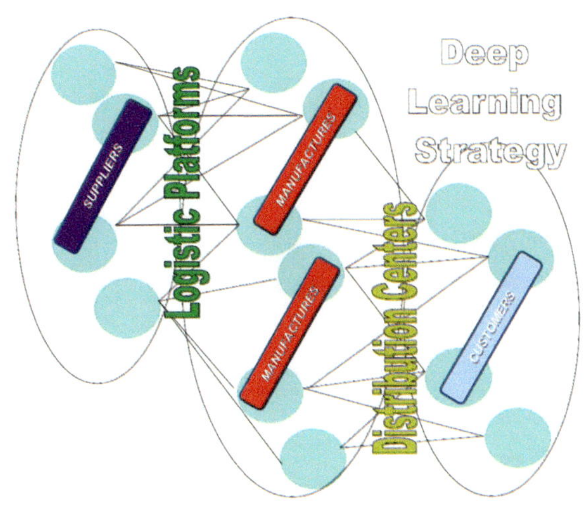

FIGURE 6.10 The multinodal strategic management of profound learning as an extended version (exogenous by "beyond itself") through the (self) interactive networking strategy of Figure 6.3; we can notice the "producers' layer" interposed to the specific layers for suppliers and customers, through interfaces containing the logistic layer of intercommunication and (re)distribution of value (goods and services); this way the feedback and a dynamic filter is providing, along the multilayered, and learning complexity. The model helps in formulating predictions and advancing proactive attitudes in developing competitive, sustainable, and regenerative businesses, with a depth learning architecture (recycled), namely, deep learning machine, passing from artificial to natural and regaining the sustainable (econo-ecolo) business space.

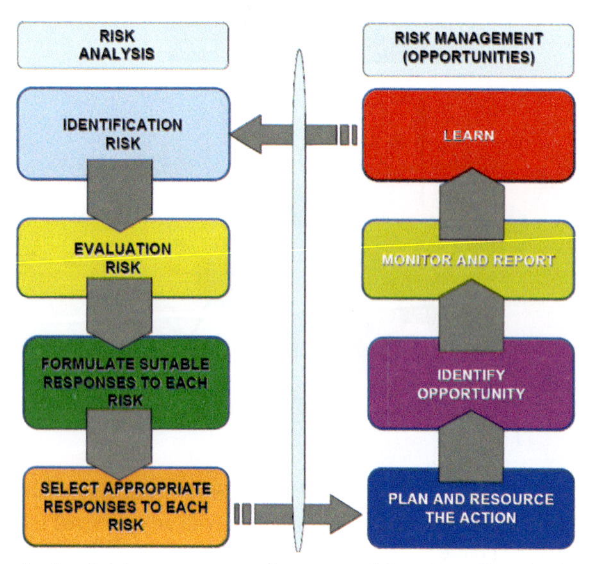

FIGURE 7.1 Cycle of risk management for successful projects [inspired and generalized according to Hristova et al. (2014, 2015)].

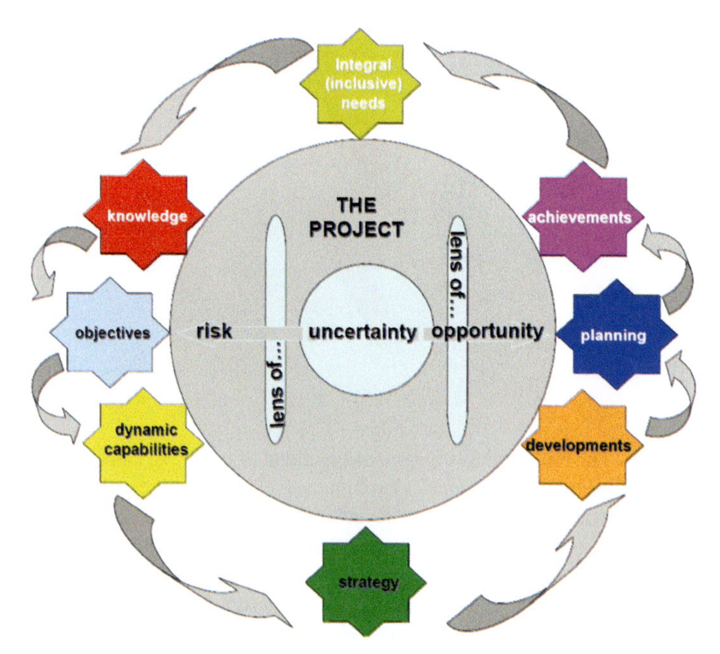

FIGURE 7.2 The role of risk and opportunity lenses to transform uncertainty into a successful project, following specifically the risk management cycle in Figure 7.1 [inspired and generalized by Chapman and Ward (2011)].

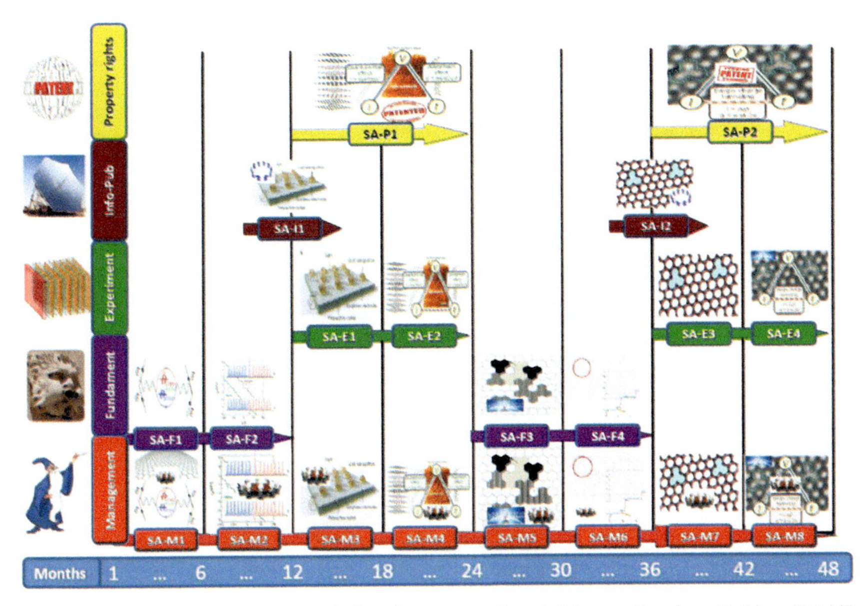

FIGURE 7.5 The structure and development of activities and sub-activities ("SA") breakdown structure and flow of NACOGRAF-2C project.

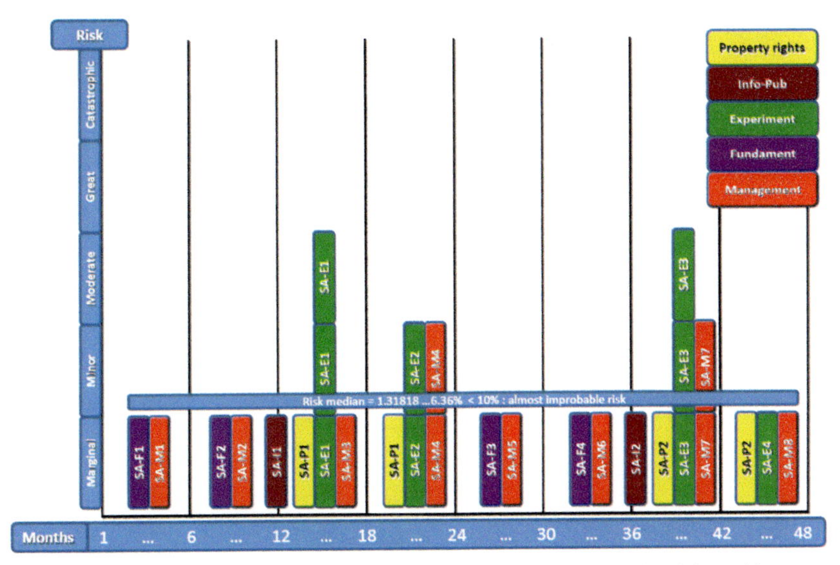

FIGURE 7.6 The graph of risk dynamics for project NACOGRAF-2C as histogram based on the risk analysis carried out in Table 7.4.

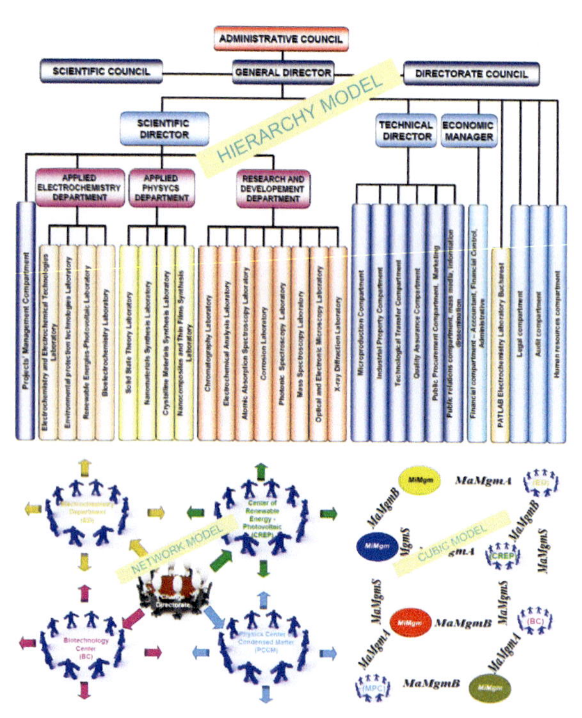

FIGURE 11.2 Top: hierarchical structure, after Drucker (1954); bottom-left: dual model network structure, after Kotter (2012); bottom-right: 3D model Cubic Strategic structure, after Putz (2017)—as the present inclusive scientific change paradigm.

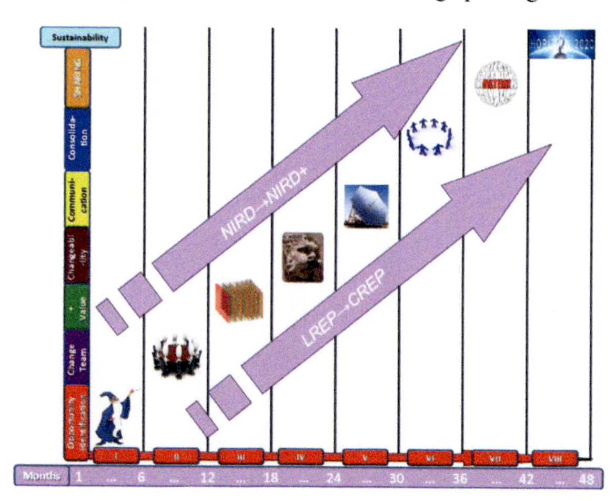

FIGURE 11.3 Gantt chart of administrative measures implementation identified from Table 11.2 in scientific inclusive change management on NIRD, Timişoara in 4 years (48 months) mandate.

CHAPTER 8

Clustering in and out Strategies of the Prisoner Dilemma in the Cube of Distinctive Advantage

ABSTRACT

The necessity of organizing the effective and efficient economic activities in business agglomerations, or within the clusters is justified (1) by the historical and conceptual premises at the industrial districts level (especially the Italian economy); (2) by the integrated business adaptation at the technological changes dynamics, based on the dynamic systems theory; (3) by the strategic games with bifurcation tension—for example, of the prisoner dilemma between two main actors of the cluster (the first being the contractor and the other the supplier system/chain—with the latest being the producers of the semi-finished products from the cluster); and (4) by the strategic cube of the distinctive advantage providing systematic evolutions in resolving the prisoner dilemma (in clusters). In all the cases, the social innovative phase of the mutual collaboration (passive phase), of getting out of contingent (proactive phase) by the stakeholders' consensus (reactive phase), is solved through the promotion of the repetitive dialog (so defining the iterative phase). The optimal solution shows up in the intermittent technological transfer and of the competition by diversification on the cluster as a whole (not just of the firms from the supplier chain).

Motto:

It is probably more difficult to be a good family father than a professional revolutionary!

—Nelson Mandela

8.1 INTRODUCTION

The synergy of business skills is a central theme in strategic management. It inevitably leads to the concept of clustering; this means a nonchaotic agglomeration of economic units, with ordered and localized (in situ) of ideas, skills, and products along with an inner (sub)chain of specific values. They regard the business units as components, either at the level of constituent firms, or the economic sectors of the organization economic orientation. The regional economy is also concerned toward a global releasing a product of superior value, while developing an alternative (often superior) advantage for all the component companies involved, respecting with their separate performance, if any.

Thus, in an increasingly globalized world, at first, there was *the industrial district!*

The industrial district, complex concept—predecessor of the post-modern cluster represents (at the same time):

- A geographical concentration of firms capable of fostering intense and frequent relationships based on mutual knowledge and *through informal (tacit) communication channels developed and consolidated over time* (Becattini, 1978, 1989, 1998)
- A complex network of geographically concentrated companies characterized by *the social division of labor* (Scott, 1982)
- A cluster of *small and medium-sized firms* that develop specialized and highly coordinated activities, necessary to produce a product (Heyter, 1997)
- In the context of resource-based strategy of firms, *resource-based view* (Wernerfelt, 1984; Amit and Schoemaker, 1993), represents *the competitive advantage of resources and competences dispersed within a multi-firm structure and nonreducible to a component firm* (Marafioti et al., 2010)
- *A crowd-sourcing concentrated in the value chain*[1] of a single business dispersed between the various activities of the small- and medium-sized firms involved; this implies that the business activity is achieved through a hierarchical (*top-down*) coordination and plan-

[1] *The value chain* in English (*catena del valore* in Italian) is a succession of activities that an organization (firm, cluster, etc.) operates in a specific industry to deliver a quality product/a valuable product on the market (Hristtova et al., 2014).

ning, being both dynamic and complex (through exchange of information and knowledge), and in a specific organizational language (possibly codified) developed over time (Piore and Sabel, 1984)

- *The existence of communication and co-operation codes* (cooperation + competition) based on a common business/economic interest (Petrişor, 2007); more recently, it includes an extension in social innovation[2] among rivals, for specific products and markets being circumscribed in a single formal organization (Bellandi, 1988; Cantwell, 1998)
- A spatial business organization in which *the degree of a resource depends on the degree of imitation and transferability* (knowledge and degree of technology) from one part (central, as predecessor of clusters) to the partner parties (Prahalad and Hamel, 1990).

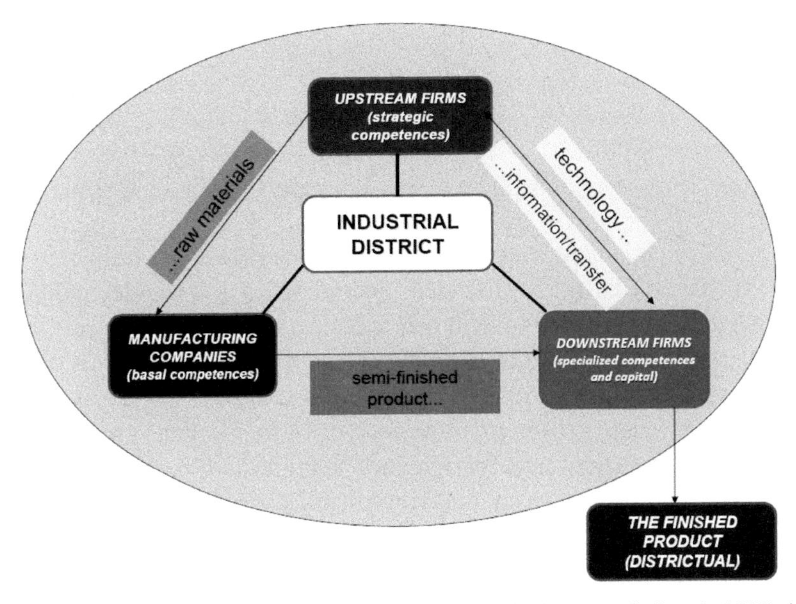

FIGURE 8.1 Relationships at the level of an industrial district (Marafioti et al., 2010); in a triptych paradigm (potential–strategy–dynamics) of a strategic management (Petrişor, 2007).

[2]*Social innovation* is all the new ideas (products, services, and models) that satisfy simultaneously (and possibly synergistically) the social needs (security, health, food, energy, transport, climate, inclusiveness) much better (more effectively) than others (ideas, old, or current paradigms) and create new relationships and societal collaborations (between social partners, including the environment, in a global view of stakeholders), see Murray et al. (2010) and Herrera (2015).

Of course, in the functioning of the industrial district, there are some fundamental issues, especially related to the transfer of knowledge, such as:

- Is it about knowledge concentrated (in some companies, or may be centrally) or is it a *uniform distribution* in an industrial district?
- And if the distribution of knowledge is not homogeneous, what kind (which is the *typology)* of firms among those involved in the chain of district value?

The answer to the first question is that knowledge distribution is not homogeneous and, moreover, it is convergent in a central firm (anticipating the form of organization as a *cluster*. In addition, the *top-down* technology transfer occurs, so closing a loop of knowledge in the cluster in order to increase the competitive advantage of global cluster business, yet facing with the specifically related risks, see next section of the cognitive analysis).

The answer to the second question, in the conditions of the nonhomogeneous distribution of knowledge (and competences) transferred to the industrial district, is synthesized in Figure 8.1, and it is systematically explained as follows:

Thus, the potential triptych of the industrial district level is as follows:

- *Manufacturing companies* (*imprese artigianale*, in Italian) provide basic skills (even artisanal, according to a possible tradition of the place—geographical and specific folk culture—social), with a role in ensuring the *distinctive potential (competitive advantage)* embedded in the finished product (final at the district level).
- *Downstream firms* (*imprese industriale*, in Italian) provide capital competencies and of specialized integration of semifinished products, as offered by manufactory firms, while delivering the finished (final) product to the market; it is the most dynamic district node in the sense that it takes semifinished raw material (*semi-lavorata*, in Italian) from the manufacturing companies in the territory. However, on the basis of information and technology exchanged with the upstream company (which provides district strategy), they deliver the finished district product; thus, downstream firms *correspond to the dynamic node* in the potential triptych of an industrial district and ensures, by its complex operation, *the sustainable advantage* of the district final product.

- *The upstream firm* has a strategic role in supplying raw material to manufacturing firms and the technology to downstream firms, depending on the information they receive from them about the characteristics of the products (possibly demanded) by the customers and market, and also about the specificity of the raw material to be provided. They ensure the compatibility with the downstream technology; so the upstream company has *a strategic role* and ensures *the regenerative advantage* in the industrial district.

Examples of such industrial districts in the footwear industry, "made in Italy" are shown in Table 8.1.

Finally, there is the question of the intra- and interdistrict interactions, schematized in Figure 8.2. Here, we can see:

- Commercial relations also occur within the districts, between local semi-finished products producers (suppliers) and finished products (final producers) for customers; it is noted the reality of *transaction costs (transfers)* reduced within the districts, on the line of (direct) *collaboration* between firms located within the district;
- Rivalry relations (the pressure) create competition in the artisan/manufacturing firms (producing the semifinished products) as operating in different districts (yet competing economically and geographically, see Table 8.1); hence, the behavior of the newly located manufacturers of finished products in relation to customers (say District no. 2 in Fig. 8.2) is to invest in technology transfer at the level of firms delivering semi-finished products in their own district; in turn, this behavior induces a competitive pressure at the level of firms of the same type in a neighboring competitor district (say District no. 1 in Fig. 8.2)—this one having to invest (even alone) in the necessary technology; the answer is possible by *following the differentiation strategy for the semi-finite product(s) at the level of District no. 1*; this way, the reaction to competition (at the same level of production in District no. 2) is manifested by sending the information of the new semi-finished product to the final producer in District no. 2; in order to avoid competition and loss of its own clients, the diversification of the customer market from District no. 1, has to co-operate with the local producer of finished products (similar

TABLE 8.1 Districts in the Footwear Industry in Italy (Marafioti et al., 2010)—for the Latest Data see "Osservatorio Nazionale Distretti Italiani."[1]

District name	Region	Specialty	No. of firms included	No. of specialized positions	Turnover (mil. Euro)	Export from production (%)
Montebelluna	Region: Veneto; Province: Treviso	Sports footwear	386	7609	1378	73
Riviera del Brenta	Region: Veneto; Province: Venezia	Luxury female	700	12,000	1900	90
Vigevano	Region: Lombardia; Province: Padova	Footwesr for mechanic industry	722	7057	250	60–80
Parabiagio	Region: Lombardia; Province: Milano	Luxury footwear	80	1500	175	–
S. Croce sull'Arno	Region: Toscana; Province: Pisa	Genuine leather footwear	800	10,000	1750	40
Valdinievole-Leporecchia	Region: Toscana; Province: Pistoia	High quality genuine leather footwear	650	3900	300	50
Macerata-Fermo	Region: Marche; Province: Macerata, Fermo	Medium and high quality footwear and components	4060	32,000	1500	65
Fusignano-Bagnacavallo	Region: Emilia-Romagna; Province: Ravenna	Economic and quality footwear	80	840	50	25
S. Mauro Pascoli	Region: Emilia-Romagna; Province: Forli-Cesena	Female luxury footwear	267	3932	12	83

[1]http://www.osservatoriodistretti.org/

TABLE 8.1 *(Continued)*

District name	Region	Specialty	No. of firms included	No. of specialized positions	Turnover (mil. Euro)	Export from production (%)
Val Vibrata	Region: Abruzzo; Province: Teramo	Footwear, shoe soles and accessories	40	<1000	50	Limited
Barletta	Region: Apulia; Province: Barletta-Andria-Trani	Sports and casual footwear	352	12,000	Ca. 7% of the national business volume	
Casarano	Region: Puglia; Province: Lecce	Sports and relaxation footwear	197	6000	500 million Euros of export	
Regiunea Napoli	Region: Campagna; Province: Napoli	Female footwear and medium-fine	143	1377	2.6% for export	

to Fig. 8.2) in District no. 2; thus, a coopetition relationship is created (competition and collaboration) on a common market of interdistrict products.

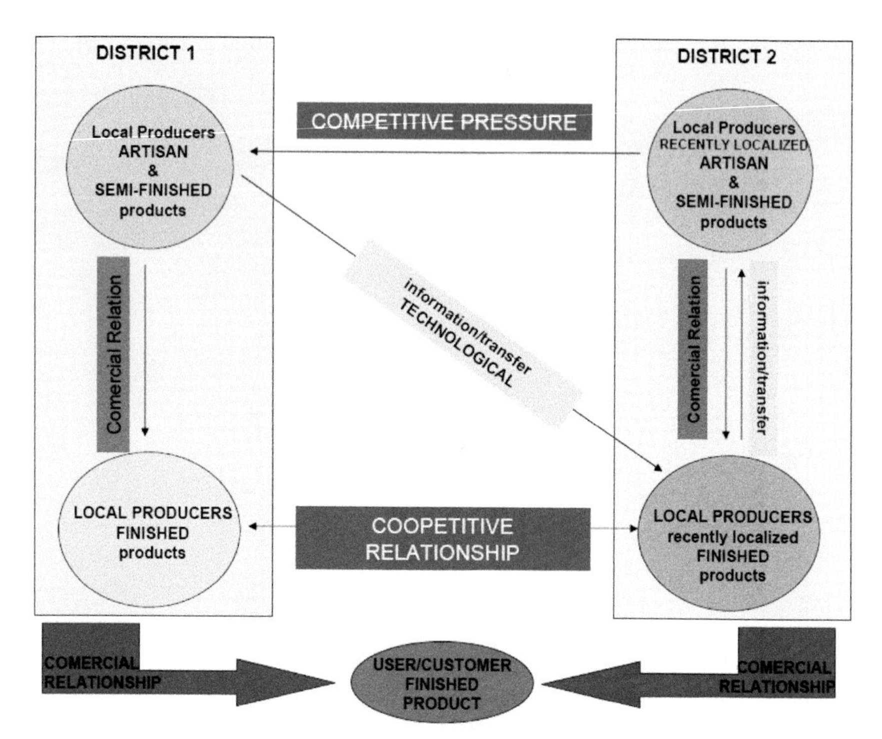

FIGURE 8.2 Economic relationships (commercial) and strategic (of competition, coopetition) intra- and interdistrict.

Source: Adapted with permission from Marafioti et al. (2010).

Two essential aspects results from this intra- and interdistrict schemes as follows:

- *The need for an interdistrict conglomeration*, intra- and geographic inter-regions (most often in the vicinity), which creates the premises, necessarily—by virtue of cooption, *for the emergence of an industrial cluster* (Porter, 1990, 1998, 2000, 2008);
- Maintaining the commercial "tension" of the competitive, sustainable, and regenerative advantage at the level of interdistrict actors, for such a cluster, relates to the transfer of technology and the

differentiation of supplying products (manufacturing and downstream, see Fig. 8.1) for the customer/market portfolio. It is based on the technological information received from the upstream firm (Fig. 8.1); so the diversification alone implemented in the manufacturing firms and downstream (Fig. 8.1)—enhances the portfolio of clients and market. The technological information received from the upstream firm of Figure 8.1 makes it become the strategic contractor in a cluster; therefore, the collaborative and the non-collaborative strategies are emerging between the cluster's main contractor and the cluster supply chain companies in a strategic game specific to the prisoner's dilemma (McAdams, 2014).

This study presents the model of dynamic systems (in the *Cognitive Analysis Section*) and then that of the distinctive advantage cube (in the *Methodological Analysis Section*) in modeling cooperative strategies of the prisoner's dilemma type.

8.2 COGNITIVE ANALYSIS

Here, the economic cluster is seen as the cooperative interaction between its *center* (*central node*), or the main contractor, and the system of suppliers. It may appear as a stage of organization subsequent to the development of productive industrial districts (with plus addedvalue to the final product), while operating on basis of the following essential subconcepts (possibly constituent):

- *The supply chain*: the chain of suppliers of semifinished products within the cluster (Sahin and Robinson, 2005; Tatikonda and Stock, 2003).
- *The system dynamics*: refers to the dynamic system of the semifinished product suppliers with the main contractor (the finite product manufacturer) comprising the economic-financial dynamics (profit) and organizational dynamics (the number of companies participating in the cluster), depending on the influence of the technological information and contracting power in a competitive advantage; it may display the inertia of changing both of the contractor and the supplier system (Forrester, 1961; Hannan and Freeman, 1977; Sterman, 2000).

- *Territorial innovation concept*: the concept of territorial innovation, in close connection with the evolution of the district in the cluster, see the Introductory Section (Moulaert and Sekia, 2003).
- *Learning regions*: regions that learn (Lawson and Lorenz 1998), related to the dynamics in clustering and delusion (Antonelli, 1995; Lundgren, 1995), see also the dynamic system above, and further exposure.
- *Flexible specialization*: flexible specialization, at the supply chain level, supported or not by the technology transfer of the main contractor in the cluster (Piore and Sabel, 1984).
- *Face-to-face:* face-to-face, related to the most effective communication between both the cluster supplier companies and the main contractor, especially in the co-operation related to information transfer, technology, and equipment (Meeus et al., 2000).
- *Commodity*: profit-making comfort, advantage in the sense of competitive and sustainable advantage, including the transfer of knowledge, skills, technology, and also low transaction cost at the cluster level, especially between the supplier system and the main contractor (Lundvall, 1992).
- *Absorptive capacity*: refers to the absorption capacity, as the ability of the cluster and its subsystems (suppliers and contractor) to perceive, understand and assimilate new information while pursuing new strategies. This behavior puts into practice the lessons learned from previous experiences, especially aimed at developing research and development activities (R&D) within it (Cohen and Levinthal, 1990) toward *sustainable productivity, but especially for sustainable productivity growth* (Porter, 1990).
- *Learning by cooperating*: learning through cooperation, in particular by pooling knowledge at the cluster level (suppliers and contractor), through administrative procedures and procedures at cluster organization level (Grant, 1996).
- *Continuum*: continuity, with reference to renewability (the regenerative advantage) in producing innovative ideas at the cluster level (Handfield and Bechtel, 2002), and also to the complementarity of semifinished products in the supply chain.
- *Outsourcing*: here, it refers to the acquisition (also the absorption) of critical value resources, as based on the external collaborations of the cluster; it appears on both parts, of the suppliers and the

contractor, which can later make the difference in products and diversification into the market; it is a strategy to create the competitive advantage (Ohmae, 1989), even by widening the cluster, from local to global, with the reference to Glocal (Robertson, 1994).

- *Revamping*: refurbishing, repairing, and also redefining—adapting to the new—especially for cluster supplier equipment, through existing retechnology investments on the existing resource basis.
- *Upgrading*: adapt to the new, with higher costs than revamping when made on the basis of technology transfer, and may include some radical changes in the production chain at suppliers, or (in the chain of processing and assembling) at the main contractor.
- *Psychological smoothing*: psychological diversification (Forrester, 1961) refers to the decision of suppliers to change their own strategy toward product differentiation and market diversification (Sterman, 1987; Sastry, 1997), outside a cluster (cluster-out);
- *Break-even*: implies the breaking of the symmetry—psychological and economic. It can be manifested by identifying the momentum of productivity growth at the level of the suppliers, and in relation to the cooperation relationship. At the behavioral level, it is a *psychological strategy,* homogeneous initially in the entire supply chain, and *economic* at the point where, the production and investment costs are equaled by income and profit to date. It acts between suppliers (in the first instance where the contractor contracting power increases within the cluster) and also at the contractor (in which case the turnover of the suppliers' increases and the transaction costs also increase at the contractor), see Lant (1992) and Schneider (1992).
- *Start-up versus spin-off*: cluster-based business development from the primary counterpart (as in a business accelerator, with certain investment capital) versus cluster business development from a vendor in the supply chain who, on the basis of the technology implemented with the support of the contractor, leaves it (stops collaboration), and follows its own business course, product, market portfolio, etc.
- *Sunk costs*: are the costs incurred "once," unrecoverable, retrospective, unlike the fixed ones (e.g., monthly, anticipated, and perspective); they appear with the implementation of technology transfer from the contractor to suppliers; if both cluster actors contribute

to the costs of investing in technology, especially information and communication, we can speak about mutual commitments in technology costs (*mutual sunk-cost commitments*), see Kim and Mahoney (2006).

- *Knowledge*: knowledge, unlike knowledge that means information (even if processed), knowledge involves a knowledge articulation in a dynamic, adjustable approach, where the accumulated experience is cognitively integrated into the current and future plans of the firm, organization, and cluster.

In this context, the main strategies in a cluster with the main contractor and a supply chain, with and without technology transfer, may show the strategic production cycles in Figure 8.3, with the respective interpretations.

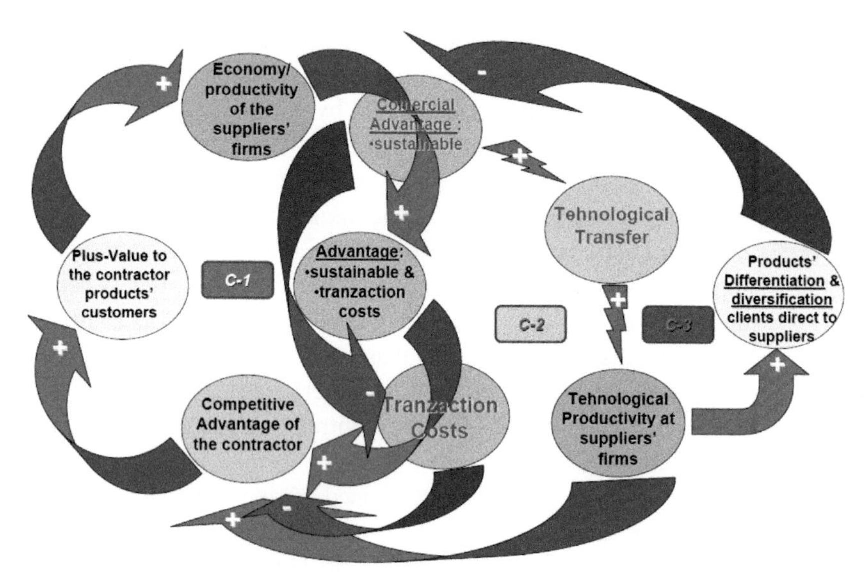

FIGURE 8.3 Feedback relationships in a cluster consisting of the main contractor and its supply chain; it includes strategic expansions when the technology transfer occurs from the first to the last; the positive (+) and negative (−) effects in clustering relationships in and out are also marked according to the adopted strategic cycle (C-1, C-2, or C-3).

Source: Adapted with permission from Antonelli et al. (2010).

- *C-1 strategic cycle*: the lead contractor has the power (or, exclusivity) to contract with cluster suppliers, whereby competitive and sustainable advantage is ensured at advantageous transaction costs within the cluster. Hence, the added value of the products reaches the cluster's customers and strengthens the economy, productivity, and productivity growth in the supply chain companies. The cycle is becoming functional (at least for a while), without additional investment in a new technology on the value chain within the cluster

- *C-2 strategic cycle*: the lead contractor decides to invest in upgrading to a new technology at the level of the suppliers in the chain; at this moment, the contractor maintains its sustainable advantage through the cluster-based contract power. Yet, it loses the advantage of low transaction costs due to nonrefundable investments in technology (namely. the sunk costs) on the vertical top-down in the cluster. The firms appreciate this type of collaboration through technology improvement, from the main contractor, so they remain faithful to it, demonstrating in turn a collaborative strategy: they return to the contractor the products waited for, in the cluster *(cluster-in)*. The improvement is given from incorporating the new technology implemented, thus enhancing the competitive advantage of the lead contractor; It nevertheless continues from here on cycle C-1
- *C-3 strategic cycle*: Part C-2 is partially resumed, with the technology investment of the main contractor on the value chain at supplier level. However, the difference is that the vendors, as medium-generic behavior, can use the new technology in order to differentiate their own products, for customer diversification outside the cluster, too *(cluster-out)*. Such behavior leads to the weakening of contracting power of the main contractor within the cluster; consequently, it loses some of its sustainable advantage, so that its transaction costs are affected within the cluster and, finally, the diminishing of the competitive advantage at the cluster level occurs; it is this the moment when the reason for clustering itself is questioned and a *bifurcation point* occurs: the lead contractor may choose to continue to collaborate and to exchange technological information with cluster providers/suppliers; even if suppliers are also diversified for other clients than those of the main contractor, they should accept some loss, by virtue of the estimates that they might lose more if stopping collaborating within the cluster owning companies with diverse clientele; alternatively, they may decide to end this collaboration at the risk of dismantling the cluster.

This is why the *analysis of collaborative–non-collaborative strategies* becomes crucial; they are part of the games theory, and they are based on the theory of dynamic systems, further exposed.

From the point of view of dynamic systems modeling (Forrester, 1961; Hannan and Freeman, 1977; Sterman, 2000), the complex relationships in the cluster as of Figure 8.3, with technology transfer, may use the parameters of Table 8.2 and be expressed according to eqs 8.1–8.5, see Antonelli et al. (2010).

TABLE 8.2 Parameters of the Dynamic System Model Corresponding to the in-and out-Clustering Cycles of Figure 8.3, for Solving Eqs 8.1–8.5.

Symbol	Meaning	Measure unit
α	Average production capacity	(Hundreds of) millions of EUR
β	Production capacity used *breakeven*	%, possibly 100%
γ	Cost of supplying firms within the suppliers chain out of the total turnover	%, possibly 30%
ε	Value of total market for suppliers (*in*- and *out*-cluster)	(Thousands) of millions of EUR
$\tilde{t}$	Technical level possible to achieve	1
τ^d	Inertia toward diversifying clientele at the cluster suppliers.	Months, possibly 12 months
τ^s	Inertia toward geographical change of suppliers (in- and out-cluster)	Months, possibly 12 months
τ^c	Inertia toward adjusting to the turnover of the main contractor	Months, possibly 6 months
τ^t	Inertia toward technological change	Months, possibly 6 months
C_0	Initial level of turnover of the main contractor	(hundreds of) million EUR
S_0	Initial level of the firms in the supply	number (dozens) of firms
T_0	Initial level of technological content	between 0 and 1
D_0	Initial level of the diversification degree amongst the suppliers	0

Source: Adapted with permission from Antonelli et al. (2010).

- *Competitive advantage*:

$$\aleph = p \cdot \phi + t \cdot (1 - \phi) \tag{8.1a}$$

is expressed, with the weighting factor $\phi \in [0,1]$, according to *the contractual power*

$$p = \begin{cases} \frac{\tilde{b}}{v} S \cdot (1-D) \ldots \frac{\tilde{b}}{v} S \cdot (1-D) > 1 \\ 1 \qquad\qquad\qquad \ldots \frac{\tilde{b}}{v} S \cdot (1-D) < 1 \end{cases} \tag{8.1b}$$

and the technical level existing at the suppliers level

$$t = 1 + \left(1 - t^g\right) \tag{8.1c}$$

In the relationship of the contracting power (eq 8.1b) it interferes: *development cost the break-even*—estimated as the product between the (average) productive capacity (α) and the productive capacity actually used for the break-even (β),

$$\tilde{b} = \alpha \cdot \beta \tag{8.1d}$$

and also the value of the *production cost of the main contractor* (v) seen as the product between the contractor's turnover (C) corrected with the contribution for the supply of the cluster suppliers (γ)

$$v = C \cdot \gamma \tag{8.1e}$$

These variables act as a ratio in the adjustment of the contractual power (8.1c) and, in turn, regulate the variable of the suppliers population (S) that did not diversify (1-D), encompassing the cluster companies remaining faithful to the main contractor, thus in collaboration with it.

On the other hand, in relation (eq 8.1c) the discrepancy factor (t^g) occurs between the current technological level (T) at supplier level and the most advanced (possibly enough) technology in the market ($\tilde{t}$):

$$t^g = 1 - \frac{T}{\tilde{t}}. \tag{8.1f}$$

This variable intervenes in formulating the dynamics of the level of technology adopted into the cluster concerned (main contractor + supplier chain) by the differential equation for:

- *Adoption dynamics of the most advanced technology in the market:*

$$\dot{T} = \frac{t^g \cdot t^D}{\tau^t}. \tag{8.2a}$$

In eq 8.2a, the level of technology desired by customers is resulted at the expense of the diversification of the clientele's portfolio of suppliers (out of cluster, *cluster-out*); this, because as a descending function of diversifying the customer portfolio, the greater the diversity achieved, the lower the distance to the technological level expected in the diversified portfolio, that is,

$$t^D = f^t(D),\ \partial f^t < 0 \tag{8.2b}$$

for products with absolute technological discrepancy (eq 8.1f), and standardized to the inertia factor in adopting/implementing any type of technology (τ^t).

- *Turnover dynamics of the main contractor:*

Expresses the deviation from the current turnover (C) due to the competitive advantage (eq 8.1a) and the inertia in achieving the current turnover (time until signing contracts with the cluster suppliers and outside cluster customers):

$$\dot{C} = \frac{C \cdot \aleph - C}{\tau^c} \tag{8.3}$$

- *Dynamics of diversification level of cluster's providers (suppliers):*

Follows the dynamics of technology adoption (eq 8.2a):

$$\dot{D} = \frac{d^S - D}{\tau^d}, \tag{8.4a}$$

where, as above, the equation registers the deviation of the current diversification versus the opportunity for diversification

$$d^S = f^d(t^g),\ \partial f^d < 0 \tag{8.4b}$$

In turn, a downward trend function with the increasing discrepancy in supplier technology over the maximum possible market technology (t^g), that is, the greater the absolute technological discrepancy gets, the lower becomes the diversity of out-cluster customers that can be served).

- *Dynamics of the cluster supplier population*:

This type of dynamics follows the same formulation as above, yet involving the difference from the current value of the population of suppliers (S)

respecting the component obtained on the contractor's production cost contributions, adapted to the value of the market approached by standard diversification to the break-even development; everything is once again standardized (normalized) to the (temporal) inertia of the change of suppliers population:

$$\dot{S} = \frac{\frac{v+\varepsilon \cdot D}{\tilde{b}} - S}{\tau^{s}} \tag{8.5}$$

From the analysis of the eqs 8.1–8.5, we can see how the cluster's dynamics is complex and intra-connected, especially with contractor (through its turnover), eventually driving the diversification of the supply chain portfolio to the *out-cluster* customers—in a balance with—the population variation of these suppliers by the cluster-out positioning (in the case of customer diversification and interruption of the collaboration with the main contractor). For the initial numerical values of the variables in Table 8.2 and for fixing the contractual power weight versus the technological degree of the suppliers in the competitive advantage (eq 8.1a), there are various collaborative–non-collaboration scenarios between the main contractor and the cluster suppliers, in order to maximize profit and specific competitive advantages, respectively. It ends thus, with a strategic game between the main contractor and his suppliers within the cluster, specific to game theory in general and to the Prisoner Dilemma in particular (McAdams, 2014):

- Both cluster actors can have a *dominant strategy,* a move that maximizes their own interest regardless of the other actor's strategy
- Both players lose more if they both play the dominant strategy than if they play any other strategy.

In this context, in the cognitive way, even in the absence of any numerical—quantitative assessments, the economic and social equivalences of the possible strategies can be formulated, by the couplings the collaborative and non-collaboration moves; see also Figure 8.4:

i) *The contractor cooperates* (continues the technology transfer to its suppliers) and *the suppliers collaborate* (remain loyal to the contractor under his contracting power, without diversifying his clientele) => *the result is passive, the cluster remains a center of local,* regional, even global interests (depending on its size).

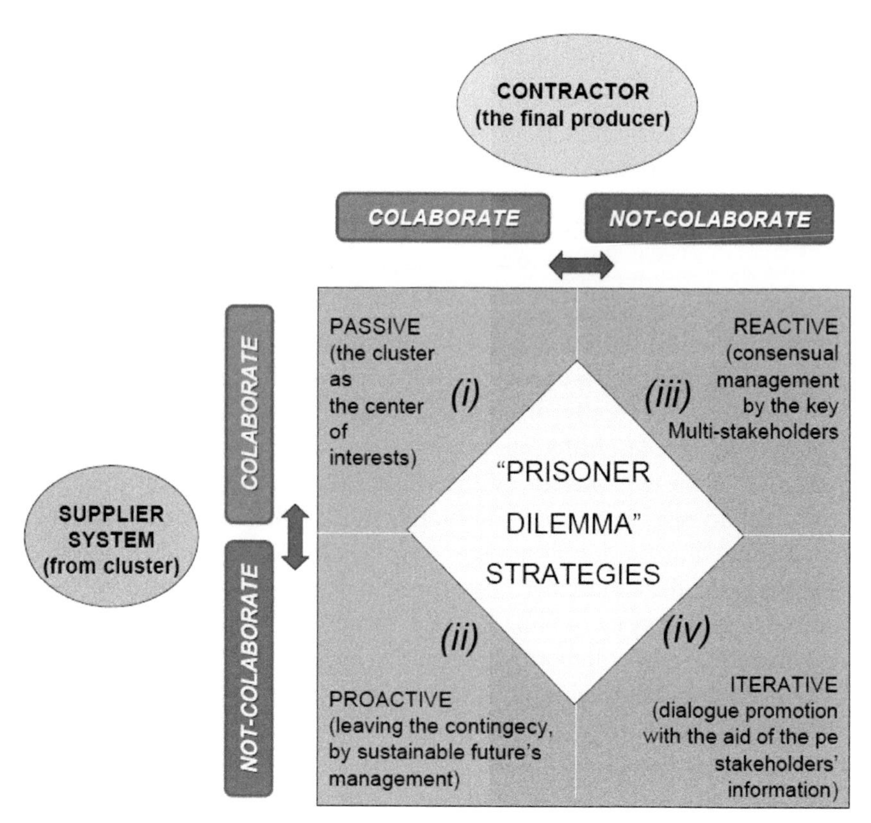

FIGURE 8.4 The generic matrix of strategic games between the final producer—the contractor and the vendor system in a cluster that can evolve with technology transfer from the first to the last behavior in the paradigm of the Prisoner Dilemma (McAdams, 2014; Petrisor, 2007); the inter-collaborative strategies are called after Herrera (2015) so giving the economic and social correct significance (in the context of social innovation) for the type of strategy adopted by the main players in the cluster.

ii) *The contractor collaborates* (continues the technology transfer to its suppliers) and *suppliers do not cooperate* (differentiates their products just on the basis of the technology transferred by the main contractor and diversifying their clientele-out of the cluster) => *the result is proactive from the side of contractor which, without shutting down the technology transfer to the current suppliers, leaves the contingency and negotiates with new vendors (future in-cluster) in order to replace the current ones (the future out-clusters).*

iii) *The contractor does not cooperate* (interrupts the technology transfer to its suppliers, possibly by the precaution of not being used in a too differentiated manner in the suppliers' products, so reducing the risk of too much cutting down the technological discrepancy from the maximum level in the market—avoiding so the consequence of favoring at the suppliers' side the differentiation of products in the first stage and then kicking them out-clustering by diversifying their own market) *and suppliers collaborate* (remain loyal to the contractor under its contracting power without diversifying its clientele) => *the result is reactive on the part of the suppliers, who will call on the key stakeholders of the cluster to persuade the main contractor to unblock the technology transfer, so facilitating the increase in the competitive advantage (of the cluster as a whole) on the market.*

iv) *Contractor does not cooperate* (interrupts technology transfer to its suppliers) and *suppliers do not cooperate* (differentiate their products even by their own investments, in order to diversify their clientele outside the cluster) => *the result is iterative, at least for a while, on both sides of the main contractor and the suppliers; repeated discussions are attempted in order to resume collaborations, while involving stakeholders and information from either sides of the local, regional and global cluster markets; paradoxically, there is the potential of enhancing the competitive advantage of both actors in the detriment of imminent de-clustering.*

It is remarkable how social innovation (see the Introductory Section) can provide an economic solution in each case of a cluster-non-collaborative, "bilingual" collaboration game. This approach is feasible for the cluster economy, perhaps exactly due to its definition structure, that is, covering local and regional interests, while using global technology. The cluster aims at cutting-edge technology, aiming at increasing productivity on an efficient basis (i.e., practicing the resource economy, with low transaction costs, excluding the sunk costs, to the detriment of efficiency (profit) at any price). Other behaviors may be "avenged" by the negative (emotional) reaction of the market, which can reject products resulting from non-collaboration in- and out-cluster. So the wise clustering should take into account all consensuses at cluster stakeholders, including the *green environment*.

A complementary, qualitative–quantitative approach is provided in the following Section 8.3, when the collaborative–non-collaborative game between the main contractor and the cluster supplier companies will be resumed with the help of the distinctive advantage cube paradigm (Putz, 2019a–2019e).

8.3 METHODOLOGICAL ANALYSIS

The triptych of the advantage competitive, sustainable, and regenerative is considered with the distinct dimensions (on 0X, 0Y, and 0Z) of a so-called "business development space: BS" (Fig. 8.5).

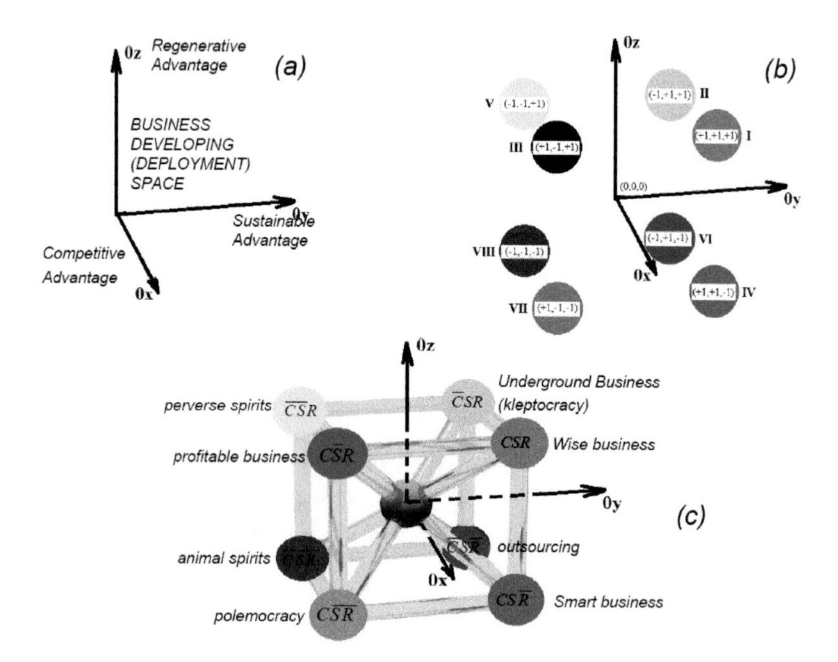

FIGURE 8.5 The business development 3D unit: (a) the three-orthogonal system developed on the axes of competitive, sustainable, and regenerative advantage, that is, on 0X, 0Y, and 0Z, respectively; (b) the eight isolated (independent) states—may correspond with the business types in the course of their life cycle development, through the synergy of the distinctive advantage, namely the (geometrically identified) positive and/or negative competitiveness, sustainability, regenerative attributes; and (c) The connection of the linking the eight types of business behaviors in a complete life-cycle development, here geometrically rooted in the coordinates obtained through combinations of competitive, sustainable and regenerative advantage/disadvantage; see also the text and Putz (2019a, 2019b).

In this space, the "unitary movements" are considered to be on any of the three axes, so only eight combinations of positive and negative coordinates of the set of values $\{-1, +1\}$ appear possible. Positive value (+1) is considered to be beneficial for that size: the company profile develops *the competitive advantage of* values +1 on the 0X axis, *the sustainable advantage of* +1 *on the 0Y axis, and the regenerative advantage over the 0Z axis*. Similarly, they can develop disadvantages for negative values on the respective BS axes. Triptych combinations indicate the complex profile of the business type, in relation to the specific advantages and disadvantages.

The next step is to quantify the interactions, the evolution, or the development of these businesses, one to the other, toward the "ultimate target"—the wise company—which manifests simultaneously (and synergistically) the competitive advantage (Petrisor, 2007), sustainable and regenerative, as (+ 1, + 1, + 1) trifecta point; The wise behavior is synergistic because, in this state, a firm is simultaneously profitable (economically competitive), sustainable (with ecological use of resources, namely, a "green" business), and regenerative (satisfies the demands of society in cycles of production, by recycling, reutilizing, reenabling, and reembedding products aiming social innovation, as defined in the Introductory Section). Quantitatively, modeling these interactions and/or business development relationships *within the business space* (BS) is done by considering the topology of these eight forms of business in a specific topology, represented by an adjacent 8×8 (A_{BS}) matrix.

$$A_{BS}^{\text{isolated}} = \begin{pmatrix} 1 & 0 & 0 & 0 & 0 & 0 & 0 & 0 \\ 0 & 1 & 0 & 0 & 0 & 0 & 0 & 0 \\ 0 & 0 & 1 & 0 & 0 & 0 & 0 & 0 \\ 0 & 0 & 0 & 1 & 0 & 0 & 0 & 0 \\ 0 & 0 & 0 & 0 & 1 & 0 & 0 & 0 \\ 0 & 0 & 0 & 0 & 0 & 1 & 0 & 0 \\ 0 & 0 & 0 & 0 & 0 & 0 & 1 & 0 \\ 0 & 0 & 0 & 0 & 0 & 0 & 0 & 1 \end{pmatrix} \tag{8.6a}$$

For example, in the form of the unit matrix: the connection between business types is absent, the situation corresponding to Figure 8.5b. Similarly, associating BS's in the direction of the *competitive, sustainable, and regenerative advantage* (Fig. 8.5c), with <+1> for positive direction raising the specific advantage, while having <−1> as opposite on the axis of the specific disadvantage of Figure 8.5a; the specific adjacent matrix is generated:

$$A_{BS}^{\text{connected } CSR} = \begin{pmatrix} 1 & -1 & -1 & -1 & 0 & 0 & 0 & 0 \\ 1 & 1 & 0 & 0 & -1 & -1 & 0 & 0 \\ 1 & 0 & 1 & 0 & -1 & 0 & -1 & 0 \\ 1 & 0 & 0 & 1 & 0 & -1 & -1 & 0 \\ 0 & 1 & 1 & 0 & 1 & 0 & 0 & -1 \\ 0 & 1 & 0 & 1 & 0 & 1 & 0 & -1 \\ 0 & 0 & 1 & 0 & 0 & 1 & 1 & -1 \\ 0 & 0 & 0 & 0 & 1 & 1 & 1 & 1 \end{pmatrix}. \tag{8.6b}$$

Under these circumstances, bifurcation cases overlap with the cases of the Prisoner Dilemma of the contractor–provider (supplier) in a cluster (Fig. 8.4). For each *strategy one unites a business type* (*as associated with the average/generic representative of the cluster supplier chain*) *from the BS to the business which has a competitive, sustainable, and regenerative advantage (associated with the main contractor in the cluster);* accordingly, the following specific matrices are considered:

i) *The Contractor COLLABORATES–the suppliers COLLABORATE*: it is considered the SHORTEST PATH linking the type of the firm (I...VII) to the supplier of type VIII (Fig. 8.5), associated with the main contractor in the cluster (the latter CONNECTED to the rest of the strategic cube).
ii) *Contractor COLLABORATES–suppliers DO NOT COLLABORATE*: it is considered the LONGEST PATH that links the type of the firm (I...VII) to type VIII supplier (Fig. 8.5), associated with the main contractor in the cluster (the latter being CONNECTED to the rest of the strategic cube).
iii) *The Contractor DOES NOT COLLABORATE–the suppliers COLLABORATE*: it is considered the SHORTEST PATH linking the type of the firm (I...VII) to type VIII supplier (Fig. 8.5), associated to the main contractor in the cluster (the latter being DISCONTINUED from the rest of the strategic cube).
iv) *The Contractor DOES NOT COLLABORATE–the suppliers DO NOT COLLABORATE*: it is considered the LONGEST PATH linking the type of the firm (I...VII) to type VIII supplier (Fig. 8.5),

associated with the main contractor in the cluster (the latter DISCONTINUED from the rest of the strategic cube).

In these cases:

- *The longest path*: associates with the lowest entropy (in suppliers) in their diversification process of clientele outside the cluster (Putz, 2019b)
- *The shortest path*: associates with direct movements in the direction of the business development space (toward suppliers), in the specific 3D order: *0X (competitiveness and competitive advantage) → 0Z (investment and regenerative advantage) → 0Y (productivity and sustainable advantage)*—a result to be re-established by a correlation study in the global knowledge economy (Putz, 2019c)
- *Disconnecting* the main contractor from the supply chain (in non-collaborative framework) is represented by the absence of the connection (i.e., put "0") in the adjacent matrix on any of the topologies linking to state VIII (a.k.a. the main contractor in the cluster), regardless the shortest path (when suppliers collaborate), or the longest path (when suppliers do not collaborate).

For each such combination, the *information index in the BS* (based on the determinant of adjacent matrix in this space), with the logistic expression (Putz, 2019d) is calculated:

$$I_{BS} = \frac{1}{\det\left[A_{BS}\right] + e^{-\det\left[A_{BS}\right]}}. \tag{8.7}$$

Analytically, for each interaction strategy (collaboration–non-collaboration) between suppliers and the main contractor in the cluster with generic technology transfer, the results are centralized in Table 8.3. The analysis in Table 8.3 leads to the following informational hierarchical strategies:

- Strategy I (I → I): PASSIVE-i <REACTIVE-iii <PROACTIVE-ii <ITERATIVE-iv
- Strategy II (II → I): PROACTIVE-ii = REACTIVE-iii = ITERATIVE-iv <PASSIVE
- Strategy III (III → I): PASSIVE-i = ITERATIVE-iv <PROACTIVE-ii = REACTIVE-iii

- Strategy IV (IV → I): PASSIVE-i = PROACTIVE-ii = REACTIVE-iii = ITERATIVE-iv
- Strategy V (V → I): PASSIVE-i = PROACTIVE-ii = REACTIVE-iii = ITERATIVE-iv
- Strategy VI (VI → I): PASSIVE-i <PROACTIVE-ii = ITERATIVE-iv <REACTIVE-iii
- Strategy VII (VII → I): REACTIVE-iii <PROACTIVE-ii = ITERATIVE-iv <PASSIVE-i
- Strategy VIII (VIII → I): PASSIVE-i = REACTIVE-iii <ITERATIVE-iv <PROACTIVE-ii

The results are particularly interesting and useful for generic solutions to the prisoner's dilemma in dynamic clusters, with acquisition and technology transfer. In this case, by virtue of the information hierarchies above (the larger the information gets, the more increased knowledge and solution finding to the prisoner's dilemma), the situations are interpreted by strategies:

- *Strategy I is not winning,* paradoxically, *it implies immobility* (I → I) in the BS—and indeed, through the absence of change, it results in the informational chain that ends in ITERATIVE, that is, in non-collaboration on both sides in the cluster, which leads to its dissolution;
- Similarly, the strategies of vendor's chains that develop strategic behavior of types IV (*smart business)* and V (*perverse spirits*) do not lead to a favorable outcome for the existence of the cluster. This happens, even if the main contractor does not react to the lack of cooperation, in the first phases, and it is passive and proactive. Observe that both approaches with cluster collaboration include the continuation of the technology transfer evolving ITERATIVELY, and finally, will dismantle the cluster.
- The strategies II (kleptocratic) and VII (polemocratic) end, both PASSIVELY, (with mutual collaboration in the cluster). Yet, they are preceded by the ITERATIVE PHASE that involves non-collaboration on both sides; so, we have here a small probability that, from simultaneous lack of collaboration, the players will go to simultaneous collaboration; consequently, in the end, this economically behavior will dismantle the cluster.

- The VI (*outsourcing*) strategy ends in the REACTIVE PHASE (say the consensus management), but it is preceded by the ITERATIVE PHASE (a very exhausting and resource consuming one, especially organizational and administrative resources in the cluster). It is therefore not a viable solution for preserving the local–global character of the system consisting of the supply chain and the main contractor. The tendency is to diversify the suppliers, while the contractor acts through outsourcing, with the results eventually dismantling of the cluster;
- Finally, the Strategy III (*profitable business*) seems to be the most appropriate attitude on the part of both cluster's actors. They use the PASSIVE phase of tacit collaboration for the ITERATIVE phase of negotiations (even through a partial or temporal break), followed by separate exploration in the PROACTIVE phase of possible markets for a de-clustered (simulated) path. Nevertheless, what followed is the consensus management succeeding in keeping the cluster in the REACTIVE phase. Maintaining the reciprocal tension is (paradoxically) providing stability (the contractor delivering from time to time new technology to the supply chain) to ensure productivity growth, whereas the latter collaborating to increase the competitive advantage of the main contractor and, overall, of the cluster in general on loop C-2 of Figure 8.3.

8.4 CONCLUSIONS AND PERSPECTIVES

The cluster is a postmodern concept of organizing economic activity, but it has its origins in the economic experience of nations, notably in terms of productivity, competitiveness, and economic growth, throughout history. The first modern forms of cluster are found in organizing, as industrial districts with the regions of Italy as eminent examples—organizational forms maintained to date. However, clusters in the postmodern organization bring to the forefront discussion an interesting hierarchical behavior. With a chain of suppliers, as small- and medium-sized firms around the main contractor–attractor, which provides the basic raw material to the component companies of the cluster, the cluster should take over the semifinished products and process them into the final product to the specific market. The organization seems to be perfect, except for the case, inevitably, of internal investments within the cluster, made by the central contractor, in

TABLE 8.3 Adjacent Matrices and Associated Information according to eq 8.7, for Bifurcations' Behaviors in the Prisoner's Dilemma in the Case of the Relationship Between the Main Contractor and the Supplier Chain at the Level of a Generic Cluster with Technology Transfer. They are based on Short–Long Paths Corresponding to Collaboration Versus Non-Cooperation on the Part of the Suppliers; see also Figure 8.4. For Each Type of Strategy, One Unites the Representative Type of Business in the Supply Chain (I....VII) with the Typology of the Main Contractor; as a Result, the Cluster Manifests the Tendency to Maximize Competitive, Sustainable, and Regenerative Advantage in State I of Figure 8.5.

Adjacent matrices and associated information in the Prisoner's Dilemma				**Specific paths for each strategy in the business space: on the longest past of the supplier (white arrows) on the shortest path (black arrows)**
Contractor COLLABORATES– Suppliers COLLABORATE (i)	**Contractor COLLABORATES– Suppliers DO NOT COLLABORATE (ii)**	**Contractor DOES NOT COLLABORATE– Suppliers COLLABORATE (iii)**	**Contractor DOES NOT COLLABORATE– Suppliers DO NOT COLLABORATE (iv)**	
Strategy I: I→I				
$\begin{pmatrix} 1 & -1 & -1 & -1 & 0 & 0 & 0 & 0 \\ 1 & 1 & 0 & 0 & -1 & -1 & 0 & 0 \\ 1 & 0 & 1 & 0 & -1 & 0 & -1 & 0 \\ 1 & 0 & 0 & 1 & 0 & -1 & -1 & 0 \\ 0 & 1 & 1 & 0 & 1 & 0 & 0 & -1 \\ 0 & 1 & 0 & 1 & 0 & 1 & 0 & -1 \\ 0 & 0 & 1 & 0 & 0 & 1 & 1 & -1 \\ 0 & 0 & 0 & 0 & 1 & 1 & 1 & 1 \end{pmatrix}$ **DET = 56** **I = 0.0178571**	$\begin{pmatrix} 1 & -1 & 0 & 1 & 0 & 0 & 0 & 0 \\ -1 & 1 & 0 & 0 & -1 & 0 & 0 & 0 \\ 0 & 0 & 1 & 0 & 1 & 0 & -1 & 0 \\ 1 & 0 & 0 & 1 & 0 & 1 & 0 & 0 \\ 0 & -1 & 1 & 0 & 1 & 0 & 0 & 0 \\ 0 & 0 & 0 & 1 & 0 & 1 & 0 & 1 \\ 0 & 0 & -1 & 0 & 0 & 0 & 1 & -1 \\ 0 & 0 & 0 & 0 & 0 & 1 & -1 & 1 \end{pmatrix}$ **DET = −3** **I = 0.058529**	$\begin{pmatrix} -1 & 0 & 0 & 0 & 0 & 0 & 0 & 0 \\ 0 & 1 & 0 & 0 & 0 & 0 & 0 & 0 \\ 0 & 0 & 1 & 0 & 0 & 0 & 0 & 0 \\ 0 & 0 & 0 & 1 & 0 & 0 & 0 & 0 \\ 0 & 0 & 0 & 0 & 1 & 0 & 0 & 0 \\ 0 & 0 & 0 & 0 & 0 & 1 & 0 & 0 \\ 0 & 0 & 0 & 0 & 0 & 0 & 1 & 0 \\ 0 & 0 & 0 & 0 & 0 & 0 & 0 & 1 \end{pmatrix}$ **DET = −1** **I = 0.581977**	$\begin{pmatrix} 1 & 0 & 0 & 0 & 0 & 0 & 0 & 0 \\ 0 & 1 & 0 & 0 & -1 & 0 & 0 & 0 \\ 0 & 0 & 1 & 0 & 1 & 0 & -1 & 0 \\ 0 & 0 & 0 & 1 & 0 & 1 & 0 & 0 \\ 0 & -1 & 1 & 0 & 1 & 0 & 0 & 0 \\ 0 & 0 & 0 & 1 & 0 & 1 & 0 & 1 \\ 0 & 0 & -1 & 0 & 0 & 0 & 1 & -1 \\ 0 & 0 & 0 & 0 & 0 & 1 & -1 & 1 \end{pmatrix}$ **DET = 1** **I = 0.731059**	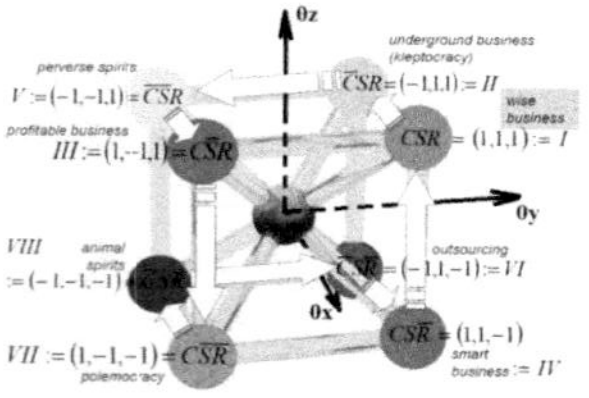
Strategy II: II→I				
$\begin{pmatrix} 1 & 1 & 0 & 0 & 0 & 0 & 0 & 0 \\ 1 & 1 & 0 & 0 & 0 & 0 & 0 & 0 \\ 0 & 0 & 1 & 0 & 0 & 0 & 0 & 0 \\ 0 & 0 & 0 & 1 & 0 & 0 & 0 & 0 \\ 0 & 0 & 0 & 0 & 1 & 0 & 0 & 0 \\ 0 & 0 & 0 & 0 & 0 & 1 & 0 & 0 \\ 0 & 0 & 0 & 0 & 0 & 0 & 1 & 0 \\ 0 & 0 & 0 & 0 & 0 & 0 & 0 & 1 \end{pmatrix}$ **DET=0** **I=1**	$\begin{pmatrix} 1 & 0 & 0 & 1 & 0 & 0 & 0 & 0 \\ 0 & 1 & 0 & 0 & -1 & 0 & 0 & 0 \\ 0 & 0 & 1 & 0 & 1 & 0 & -1 & 0 \\ 1 & 0 & 0 & 1 & 0 & 1 & 0 & 0 \\ 0 & 0 & 1 & 0 & 1 & 0 & 0 & 0 \\ 0 & 0 & 0 & 1 & 0 & 1 & 0 & 1 \\ 0 & 0 & -1 & 0 & 0 & 0 & 1 & -1 \\ 0 & 0 & 0 & 0 & 0 & 1 & -1 & 1 \end{pmatrix}$ **DET=1** **I=0.731059**	$\begin{pmatrix} 1 & 0 & 0 & 0 & 0 & 0 & 0 & 0 \\ 0 & 1 & 0 & 0 & 0 & 0 & 0 & 0 \\ 0 & 0 & 1 & 0 & 0 & 0 & 0 & 0 \\ 0 & 0 & 0 & 1 & 0 & 0 & 0 & 0 \\ 0 & 0 & 0 & 0 & 1 & 0 & 0 & 0 \\ 0 & 0 & 0 & 0 & 0 & 1 & 0 & 0 \\ 0 & 0 & 0 & 0 & 0 & 0 & 1 & 0 \\ 0 & 0 & 0 & 0 & 0 & 0 & 0 & 1 \end{pmatrix}$ **DET=1** **I=0.731059**	$\begin{pmatrix} 1 & 0 & 0 & 0 & 0 & 0 & 0 & 0 \\ 0 & 1 & 0 & 0 & -1 & 0 & 0 & 0 \\ 0 & 0 & 1 & 0 & 1 & 0 & -1 & 0 \\ 0 & 0 & 0 & 1 & 0 & 1 & 0 & 0 \\ 0 & 0 & 1 & 0 & 1 & 0 & 0 & 0 \\ 0 & 0 & 0 & 1 & 0 & 1 & 0 & 1 \\ 0 & 0 & -1 & 0 & 0 & 0 & 1 & -1 \\ 0 & 0 & 0 & 0 & 0 & 1 & -1 & 1 \end{pmatrix}$ **DET=1** **I=0.731059**	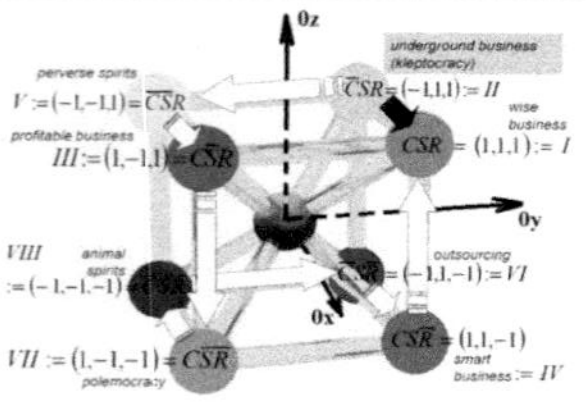

TABLE 8.3 *(Continued)*

Adjacent matrices and associated information in the Prisoner's Dilemma				Specific paths for each strategy in the business space: on the longest past of the supplier (white arrows) on the shortest path (black arrows)
Contractor COLLABORATES–Suppliers COLLABORATE (i)	**Contractor COLLABORATES–Suppliers DO NOT COLLABORATE (ii)**	**Contractor DOES NOT COLLABORATE–Suppliers COLLABORATE (iii)**	**Contractor DOES NOT COLLABORATE–Suppliers DO NOT COLLABORATE (iv)**	
Strategy III: III→I				
$\begin{pmatrix} 1&0&0&1&0&0&0&0\\ 0&1&0&0&0&0&0&0\\ 0&0&1&0&-1&0&0&0\\ 1&0&0&1&0&1&0&0\\ 0&0&-1&0&1&0&0&-1\\ 0&0&0&1&0&1&0&1\\ 0&0&0&0&0&0&1&0\\ 0&0&0&0&-1&1&0&1 \end{pmatrix}$ **DET = 1** **I = 0.731059**	$\begin{pmatrix} 1&0&0&1&0&0&0&0\\ 0&1&0&0&1&-1&0&0\\ 0&0&1&0&-1&0&0&0\\ 1&0&0&1&0&0&1&0\\ 0&1&-1&0&1&0&0&0\\ 0&-1&0&0&0&1&0&-1\\ 0&0&0&1&0&0&1&1\\ 0&0&0&0&0&-1&1&1 \end{pmatrix}$ **DET = 0** **I = 1**	$\begin{pmatrix} 1&0&0&0&0&0&0&0\\ 0&1&0&0&0&0&0&0\\ 0&0&1&0&-1&0&0&0\\ 0&0&0&1&0&1&0&0\\ 0&0&-1&0&1&0&0&-1\\ 0&0&0&1&0&1&0&1\\ 0&0&0&0&0&0&1&0\\ 0&0&0&0&-1&1&0&1 \end{pmatrix}$ **DET = 0** **I = 1**	$\begin{pmatrix} 1&0&0&0&0&0&0&0\\ 0&1&0&0&1&-1&0&0\\ 0&0&1&0&-1&0&0&0\\ 0&0&0&1&0&0&1&0\\ 0&1&-1&0&1&0&0&0\\ 0&-1&0&0&0&1&0&-1\\ 0&0&0&1&0&0&1&1\\ 0&0&0&0&0&-1&1&1 \end{pmatrix}$ **DET = 1** **I = 0.731059**	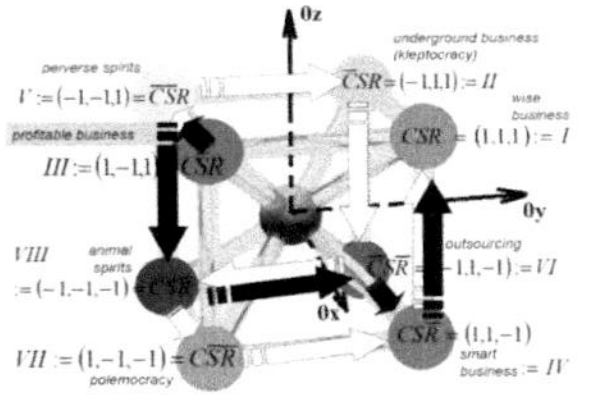
Strategy IV: IV→I				
$\begin{pmatrix} 1&-1&-1&-1&0&0&0&0\\ 1&1&0&0&-1&1&0&0\\ 1&0&1&0&1&0&-1&0\\ 1&0&0&1&0&-1&1&0\\ 0&-1&1&0&1&0&0&0\\ 0&1&0&-1&0&1&0&0\\ 0&0&0&0&0&0&1&0\\ 0&0&-1&1&0&0&0&1 \end{pmatrix}$ **DET=0** **I=1**	$\begin{pmatrix} 1&1&0&0&0&0&0&0\\ 1&1&0&0&1&0&0&0\\ 0&0&1&0&-1&0&1&0\\ 0&0&0&1&0&-1&0&0\\ 0&1&-1&0&1&0&0&0\\ 0&0&0&-1&0&1&0&-1\\ 0&0&1&0&0&0&1&1\\ 0&0&0&0&0&-1&1&1 \end{pmatrix}$ **DET=0** **I=1**	$\begin{pmatrix} 1&0&0&0&0&0&0&0\\ 0&1&0&0&-1&1&0&0\\ 0&0&1&0&1&0&-1&0\\ 0&0&0&1&0&-1&1&0\\ 0&-1&1&0&1&0&0&0\\ 0&1&0&-1&0&1&0&0\\ 0&0&0&0&0&0&1&0\\ 0&0&-1&1&0&0&0&1 \end{pmatrix}$ **DET=0** **I=1**	$\begin{pmatrix} 1&0&0&0&0&0&0&0\\ 0&1&0&0&1&0&0&0\\ 0&0&1&0&-1&0&1&0\\ 0&0&0&1&0&-1&0&0\\ 0&1&-1&0&1&0&0&0\\ 0&0&0&-1&0&1&0&-1\\ 0&0&1&0&0&0&1&1\\ 0&0&0&0&0&-1&1&1 \end{pmatrix}$ **DET=0** **I=1**	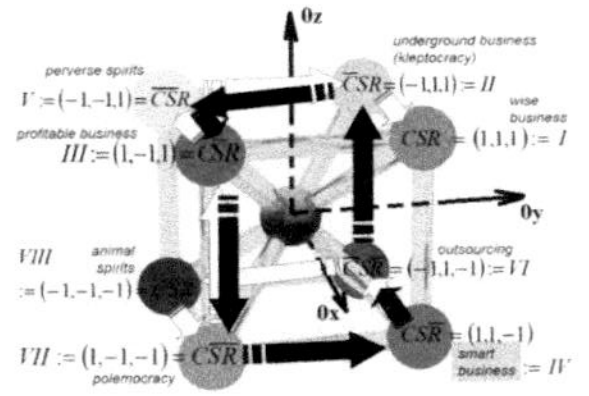

TABLE 8.3 *(Continued)*

Adjacent matrices and associated information in the Prisoner's Dilemma				Specific paths for each strategy in the business space: on the longest past of the supplier (white arrows) on the shortest path (black arrows)
Contractor COLLABORATES– Suppliers COLLABORATE (i)	**Contractor COLLABORATES– Suppliers DO NOT COLLABORATE (ii)**	**Contractor DOES NOT COLLABORATE– Suppliers COLLABORATE (iii)**	**Contractor DOES NOT COLLABORATE– Suppliers DO NOT COLLABORATE (iv)**	
		Strategy V: V→I		
$\begin{pmatrix} 1 & -1 & -1 & -1 & 0 & 0 & 0 & 0 \\ 1 & 1 & 0 & 0 & -1 & 1 & 0 & 0 \\ 1 & 0 & 1 & 0 & 1 & 0 & -1 & 0 \\ 1 & 0 & 0 & 1 & 0 & -1 & 1 & 0 \\ 0 & -1 & 1 & 0 & 1 & 0 & 0 & 0 \\ 0 & 1 & 0 & -1 & 0 & 1 & 0 & 0 \\ 0 & 0 & 0 & 0 & 0 & 0 & 1 & 0 \\ 0 & 0 & -1 & 1 & 0 & 0 & 0 & 1 \end{pmatrix}$ **DET = 0** **I = 1**	$\begin{pmatrix} 1 & 0 & 0 & 1 & 0 & 0 & 0 & 0 \\ 0 & 1 & 0 & 0 & 0 & 0 & 0 & 0 \\ 0 & 0 & 1 & 0 & 1 & 0 & -1 & 0 \\ 1 & 0 & 0 & 1 & 0 & 1 & 0 & 0 \\ 0 & 0 & 1 & 0 & 1 & 0 & 0 & 0 \\ 0 & 0 & 0 & 1 & 0 & 1 & 0 & 1 \\ 0 & 0 & -1 & 0 & 0 & 0 & 1 & -1 \\ 0 & 0 & 0 & 0 & 0 & 1 & -1 & 1 \end{pmatrix}$ **DET = 0** **I = 1**	$\begin{pmatrix} 1 & 0 & 0 & 0 & 0 & 0 & 0 & 0 \\ 0 & 1 & 0 & 0 & -1 & 1 & 0 & 0 \\ 0 & 0 & 1 & 0 & 1 & 0 & -1 & 0 \\ 0 & 0 & 0 & 1 & 0 & -1 & 1 & 0 \\ 0 & -1 & 1 & 0 & 1 & 0 & 0 & 0 \\ 0 & 1 & 0 & -1 & 0 & 1 & 0 & 0 \\ 0 & 0 & 0 & 0 & 0 & 0 & 1 & 0 \\ 0 & 0 & -1 & 1 & 0 & 0 & 0 & 1 \end{pmatrix}$ **DET = 0** **I = 1**	$\begin{pmatrix} 1 & 0 & 0 & 0 & 0 & 0 & 0 & 0 \\ 0 & 1 & 0 & 0 & 0 & 0 & 0 & 0 \\ 0 & 0 & 1 & 0 & 1 & 0 & -1 & 0 \\ 0 & 0 & 0 & 1 & 0 & 1 & 0 & 0 \\ 0 & 0 & 1 & 0 & 1 & 0 & 0 & 0 \\ 0 & 0 & 0 & 1 & 0 & 1 & 0 & 1 \\ 0 & 0 & -1 & 0 & 0 & 0 & 1 & -1 \\ 0 & 0 & 0 & 0 & 0 & 1 & -1 & 1 \end{pmatrix}$ **DET = 0** **I = 1**	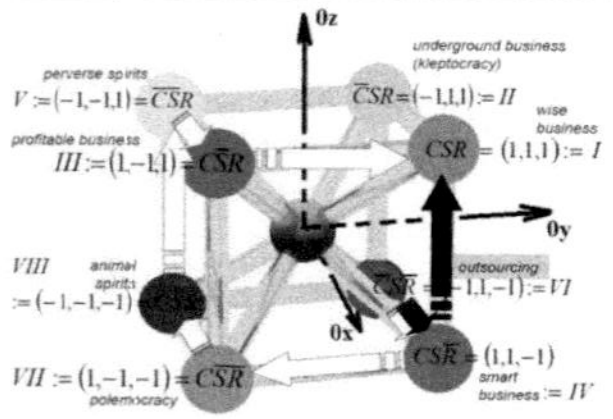
		Strategy VI: VI→I		
$\begin{pmatrix} 1 & 0 & 0 & 1 & 0 & 0 & 0 & 0 \\ 0 & 1 & 0 & 0 & 0 & 0 & 0 & 0 \\ 0 & 0 & 1 & 0 & 0 & 0 & 0 & 0 \\ 1 & 0 & 0 & 1 & 0 & 1 & 0 & 0 \\ 0 & 0 & 0 & 0 & 1 & 0 & 0 & 0 \\ 0 & 0 & 0 & 1 & 0 & 1 & 0 & 0 \\ 0 & 0 & 0 & 0 & 0 & 0 & 1 & 0 \\ 0 & 0 & 0 & 0 & 0 & 0 & 0 & 1 \end{pmatrix}$ **DET = −1** **I = 0.581977**	$\begin{pmatrix} 1 & 0 & 1 & 0 & 0 & 0 & 0 & 0 \\ 0 & 1 & 0 & 0 & 0 & 0 & 0 & 0 \\ 1 & 0 & 1 & 0 & 1 & 0 & 0 & 0 \\ 0 & 0 & 0 & 1 & 0 & 1 & -1 & 0 \\ 0 & 0 & 1 & 0 & 1 & 0 & 0 & 1 \\ 0 & 0 & 0 & 1 & 0 & 1 & 0 & 0 \\ 0 & 0 & 0 & -1 & 0 & 0 & 1 & -1 \\ 0 & 0 & 0 & 0 & 1 & 0 & -1 & 1 \end{pmatrix}$ **DET = 1** **I = 0.731059**	$\begin{pmatrix} 1 & 0 & 0 & 0 & 0 & 0 & 0 & 0 \\ 0 & 1 & 0 & 0 & 0 & 0 & 0 & 0 \\ 0 & 0 & 1 & 0 & 0 & 0 & 0 & 0 \\ 0 & 0 & 0 & 1 & 0 & 1 & 0 & 0 \\ 0 & 0 & 0 & 0 & 1 & 0 & 0 & 0 \\ 0 & 0 & 0 & 1 & 0 & 1 & 0 & 0 \\ 0 & 0 & 0 & 0 & 0 & 0 & 1 & 0 \\ 0 & 0 & 0 & 0 & 0 & 0 & 0 & 1 \end{pmatrix}$ **DET = 0** **I = 1**	$\begin{pmatrix} 1 & 0 & 0 & 0 & 0 & 0 & 0 & 0 \\ 0 & 1 & 0 & 0 & 0 & 0 & 0 & 0 \\ 0 & 0 & 1 & 0 & 1 & 0 & 0 & 0 \\ 0 & 0 & 0 & 1 & 0 & 1 & -1 & 0 \\ 0 & 0 & 1 & 0 & 1 & 0 & 0 & 1 \\ 0 & 0 & 0 & 1 & 0 & 1 & 0 & 0 \\ 0 & 0 & 0 & -1 & 0 & 0 & 1 & -1 \\ 0 & 0 & 0 & 0 & 1 & 0 & -1 & 1 \end{pmatrix}$ **DET = 1** **I = 0.731059**	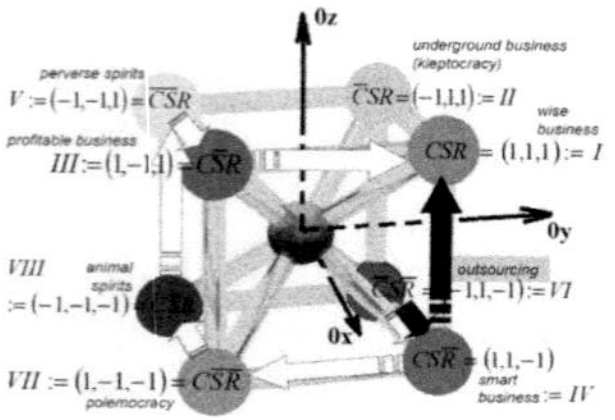

TABLE 8.3 *(Continued)*

Adjacent matrices and associated information in the Prisoner's Dilemma				Specific paths for each strategy in the business space: on the longest past of the supplier (white arrows) on the shortest path (black arrows)
Contractor COLLABORATES–Suppliers COLLABORATE (i)	Contractor COLLABORATES–Suppliers DO NOT COLLABORATE (ii)	Contractor DOES NOT COLLABORATE–Suppliers COLLABORATE (iii)	Contractor DOES NOT COLLABORATE–Suppliers DO NOT COLLABORATE (iv)	
Strategy VII: VII→I				
$\begin{pmatrix} 1 & 1 & 0 & 0 & 0 & 0 & 0 & 0 \\ 1 & 1 & 0 & 0 & 1 & 0 & 0 & 0 \\ 0 & 0 & 1 & 0 & 0 & 0 & 0 & 0 \\ 0 & 0 & 0 & 1 & 0 & 0 & 0 & 0 \\ 0 & 1 & 0 & 0 & 1 & 0 & 0 & 1 \\ 0 & 0 & 0 & 0 & 0 & 1 & 0 & 0 \\ 0 & 0 & 0 & 0 & 0 & 0 & 1 & -1 \\ 0 & 0 & 0 & 0 & 1 & 0 & -1 & 1 \end{pmatrix}$	$\begin{pmatrix} 1 & 0 & 1 & 0 & 0 & 0 & 0 & 0 \\ 0 & 1 & 0 & 0 & -1 & 1 & 0 & 0 \\ 1 & 0 & 1 & 0 & 1 & 0 & 0 & 0 \\ 0 & 0 & 0 & 1 & 0 & 0 & 0 & 0 \\ 0 & -1 & 1 & 0 & 1 & 0 & 0 & 0 \\ 0 & 1 & 0 & 0 & 0 & 1 & 0 & 1 \\ 0 & 0 & 0 & 0 & 0 & 0 & 1 & -1 \\ 0 & 0 & 0 & 0 & 0 & 1 & -1 & 1 \end{pmatrix}$	$\begin{pmatrix} 1 & 0 & 0 & 0 & 0 & 0 & 0 & 0 \\ 0 & 1 & 0 & 0 & 1 & 0 & 0 & 0 \\ 0 & 0 & 1 & 0 & 0 & 0 & 0 & 0 \\ 0 & 0 & 0 & 1 & 0 & 0 & 0 & 0 \\ 0 & 1 & 0 & 0 & 1 & 0 & 0 & 1 \\ 0 & 0 & 0 & 0 & 0 & 1 & 0 & 0 \\ 0 & 0 & 0 & 0 & 0 & 0 & 1 & -1 \\ 0 & 0 & 0 & 0 & 1 & 0 & -1 & 1 \end{pmatrix}$	$\begin{pmatrix} 1 & 0 & 0 & 0 & 0 & 0 & 0 & 0 \\ 0 & 1 & 0 & 0 & -1 & 1 & 0 & 0 \\ 0 & 0 & 1 & 0 & 1 & 0 & 0 & 0 \\ 0 & 0 & 0 & 1 & 0 & 0 & 0 & 0 \\ 0 & -1 & 1 & 0 & 1 & 0 & 0 & 0 \\ 0 & 1 & 0 & 0 & 0 & 1 & 0 & 1 \\ 0 & 0 & 0 & 0 & 0 & 0 & 1 & -1 \\ 0 & 0 & 0 & 0 & 0 & 1 & -1 & 1 \end{pmatrix}$	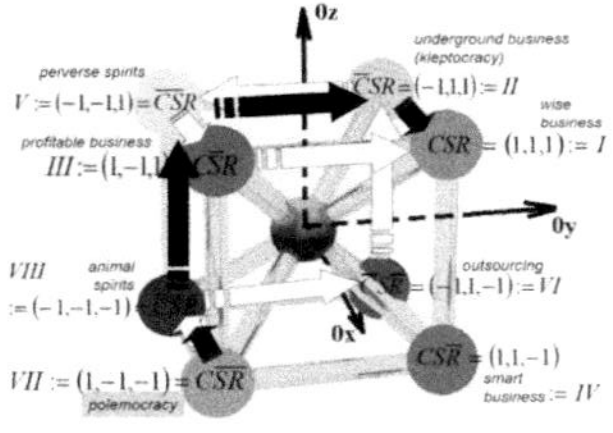
DET = 0 **I = 1**	**DET = 1** **I = 0.731059**	**DET = −1** **I = 0.581977**	**DET = 1** **I = 0.731059**	
Strategy VIII: VIII→I				
$\begin{pmatrix} 1 & 0 & 1 & 0 & 0 & 0 & 0 & 0 \\ 0 & 1 & 0 & 0 & 0 & 0 & 0 & 0 \\ 1 & 0 & 1 & 0 & 0 & 0 & 1 & 0 \\ 0 & 0 & 0 & 1 & 0 & 0 & 0 & 0 \\ 0 & 0 & 0 & 0 & 1 & 0 & 0 & 0 \\ 0 & 0 & 0 & 0 & 0 & 1 & 0 & 0 \\ 0 & 0 & 1 & 0 & 0 & 0 & 1 & 1 \\ 0 & 0 & 0 & 0 & 0 & 0 & 1 & 1 \end{pmatrix}$	$\begin{pmatrix} 1 & 0 & 1 & 0 & 0 & 0 & 0 & 0 \\ 0 & 1 & 0 & 0 & -1 & 1 & 0 & 0 \\ 1 & 0 & 1 & 0 & 1 & 0 & 0 & 0 \\ 0 & 0 & 0 & 1 & 0 & -1 & 1 & 0 \\ 0 & -1 & 1 & 0 & 1 & 0 & 0 & 0 \\ 0 & 1 & 0 & -1 & 0 & 1 & 0 & 0 \\ 0 & 0 & 0 & 1 & 0 & 0 & 1 & 1 \\ 0 & 0 & 0 & 0 & 0 & 0 & 1 & 1 \end{pmatrix}$	$\begin{pmatrix} 1 & 0 & 0 & 0 & 0 & 0 & 0 & 0 \\ 0 & 1 & 0 & 0 & 0 & 0 & 0 & 0 \\ 0 & 0 & 1 & 0 & 0 & 0 & 1 & 0 \\ 0 & 0 & 0 & 1 & 0 & 0 & 0 & 0 \\ 0 & 0 & 0 & 0 & 1 & 0 & 0 & 0 \\ 0 & 0 & 0 & 0 & 0 & 1 & 0 & 0 \\ 0 & 0 & 1 & 0 & 0 & 0 & 1 & 1 \\ 0 & 0 & 0 & 0 & 0 & 0 & 1 & 1 \end{pmatrix}$	$\begin{pmatrix} 1 & 0 & 0 & 0 & 0 & 0 & 0 & 0 \\ 0 & 1 & 0 & 0 & -1 & 1 & 0 & 0 \\ 0 & 0 & 1 & 0 & 1 & 0 & 0 & 0 \\ 0 & 0 & 0 & 1 & 0 & -1 & 1 & 0 \\ 0 & -1 & 1 & 0 & 1 & 0 & 0 & 0 \\ 0 & 1 & 0 & -1 & 0 & 1 & 0 & 0 \\ 0 & 0 & 0 & 1 & 0 & 0 & 1 & 1 \\ 0 & 0 & 0 & 0 & 0 & 0 & 1 & 1 \end{pmatrix}$	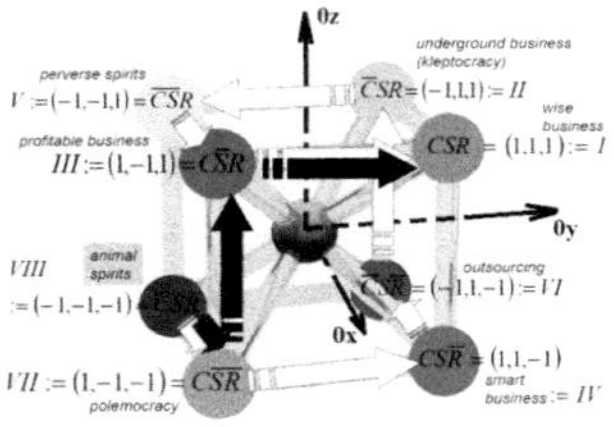
DET = −1 **I = 0.581977**	**DET = 0** **I = 1**	**DET= −-1** **I = 0.581977**	**DET = 1** **I = 0.731059**	

current and advanced technology, in order to renew and increase productivity, respectively, the competitive advantage of the cluster through technology enriched products. At this point, there is an economic behavioral bifurcation that requires strategic decisions, both at suppliers and on the cluster contractor sides: the temptation to use technology to differentiate products at the level of the suppliers in the chain, with the consequence of diversifying the market outside the main contractor and the cluster, creates clustering tension and dilemmas. In this strategic game of internal collaboration and competition, due to the weakening of the competitive advantage of the contractor, the increase of the transaction costs to its own suppliers may rise up to the danger of the cluster being dismantled. Collaboration of the contractor means continuing the technology transfer within the cluster, while the collaboration of the supply chain companies means continuing the privileged relationship with the contractor—ensuring its exclusivity in supply and preserving its contractual power, low transfer costs, etc. The situation is unlikely in the long run, even for companies with increased economic productivity, where from the prisoner's dilemma (suppliers and contractor are as "prisoners" in the cluster) linked to the optimal strategy in collaboration—competition, so that, each actor would maximize his interest (i.e., increasing the in- and out-cluster competitive advantage). The model of dynamic systems is presented in the cognitive analysis but, in the methodological part, original in this study, the solution to the strategic dilemma of competition is provided at the level of information, shared by various firms in the so-called business development space with the prime contractor, seen as the business with competitive, sustainable, and, regenerative advantage behavior. By combining the minimum–maximum information related to the path (in the sense of evolution and change of business typology) in the generated cubic space, (the cube of the distinctive advantage), various scenarios of competition are systematically obtained and, appropriately, interpreted. The equilibrium solution for the prisoners (suppliers and contractor) is achieved only in the dynamical cluster with internal technology transferred, and allowing the diversification of the market outside it. The conclusion is refined by the exclusive contractor's collaboration (i.e., performing the technology transfer to suppliers), interruptions included. Such disconnected behavior favors the pro-active prospecting; when it is performed (i.e., the diversification for cluster clientele as a whole), the collaboration may go through the consensus management working on the reactive tension (contractor–supplier's potential difference of capabilities). Thus, the cluster can be ensured with continuity of productivity through in- and out-clustering cooperation. Subsequent approaches of the couple suppliers—contractor within the cluster, on algebraic basis—are based on catastrophe theory (see also Chapter 3 of the present monograph), especially for contractor modeling as an attractor; they are possible and necessary for a multiscalar vision of the strategic potential of clusters in developing the distinctive multidimensional advantage (Putz 2017, 2019e).

KEYWORDS

- **industrial district**
- **dynamic clusters**
- **contractor and supplier**
- **game theory**
- **strategic cube of the distinctive advantage**
- **adjacency matrix**
- **information in the business space**

REFERENCES

Amit, R.; Schoemaker P. J. H. Strategic Assets and Organizational Rent. *Strateg. Manag. J.* **1993,** *14* (1), 33–46.

Antonelli, C. *The Economics of Localized Technological Change and Industrial Dynamics*. Kluwer Academic Publishers: Dordrecht/Boston/London, 1995.

Antonelli, G.; Mollona, E.; Moschera, L. Dinamiche evolutive in un cluster di produzione: una simulazione deiprocessi di interazione strategica e collaborazione. In *Dinamiche evolutive nei cluster geografici di imprese*; Boari, C., Ed.; II Mulino: Bologna, 2010.

Becattini, G. The Development of Light Industry in Tuscany: An Interpretation. *Econ. Notes*; **1978,** *2*, 107–123.

Becattini, G. Sectors and/or Districts: Some Remarks on the Conceptual Foundation of Industrial Economics. In *Small Firms and Industrial Districts in Italy*; Goodman, E., Bamford, J., Eds.; Routledge: London, 1989.

Becattini, G. *Distretti Industriali e (Made in Italy)*; Bollati Boringhieri: Torino, 1998.

Bellandi, M. The Industrial District in Marshall. In *Small Firms and Industrial Districts in Italy*; Goodman, E., Bamford, J., Eds.; Routledge: London, 1989.

Cantwell, J. Knowledge, Capabilities, Imagination and Cooperation in Business: Introduction. *J. Econ. Behav. Org.* **1998,** *35*, 133–137.

Cohen, W. M.; Levinthal, D. A. Absorptive Capacity: A New Perspective on Learning and Innovation. *Admin. Sci. Q.* **1990,** *35*, 128–152.

Forrester, J. W. *Industrial Dynamics*; The MIT Press: Cambridge, Massachusetts, 1961.

Grant, R. Prospering in Dynamically-Competitive Environments: Organizational Capability as Knowledge Integration. *Org. Sci.* **1996,** *7*, 375–387.

Handfield, R.; Bechtel, C. The Role of Trust and Relationship Structure in Improving Supply Chain Responsiveness. *Ind. Mark. Manag.* **2002,** *31* (4), 367–382.

Hannan, M. T.; Freeman, J. The Population Ecology of Organization. *Am. J. Sociol.* **1977,** *82* (5), 929–964.

Hayter, R. *The Dynamics of Industrial Location*; Wiley: New York, 1997.

Herrera, M. E. B. Creating Competitive Advantage by Institutionalizing Corporate Social Innovation. *J. Bus. Res.* **2015,** *68*, 1468–1474.

Hristova, I.; Ilic, I.; Kocjancic, A.; Struna, D.; Zajc, M. D. *Beyond the Horizon. Practical Guide to Developing Competitive Project Proposals in Horizon 2020*; RR&CO. Knowledge Centre Ltd.: Ljubljana, 2014.

Kim, S. M.; Mahoney, J. T. Mutual Commitment to Support Exchange: Relation-Specific IT System as a Substitute for Managerial Hierarchy. *Strateg. Manag. J.* **2006,** *27*, 401–423.

Lant, T. K. Aspiration Level Adaptation: An Empirical Exploration. *Manag. Sci.* **1992,** *38*, 623–644.

Lawson, C.; Lorentz, E. Collective Learning, Tacit Knowledge and Regional Innovative Capacity. *Reg. Stud.* **1998,** *33*, 305–317.

Lundgren, A. *Technological Innovation and Network Evolution;* Routledge: London, 1995.

Lundvall, B. Å. User–Producer Relationships, National Systems of Innovation and Internationalization. In *National Sytems of Innovation: Towards a Theory of Innovation and Interactive Learning*; Lundvall, B. Å., Ed.; Pinter Publishers: London, 1992.

Marafioti, E.; Mollona, E.; Perretti, F. Internazionalizzazione e competitività: una teoria dinamica dell'evoluzione dei distretti italiani. In *Dinamiche evolutive nei cluster geografici di imprese*; Boari, C. Ed.; Il Mulino: Bologna, 2010.

McAdams, D. *Game-Changer: Game Theory and the Art of Transforming Strategic Situations*; W. W. Norton & Company, Inc.: New York, 2014.

Meeus, M.; Oerlemans, L.; Van Dijck, J. Interactive Learning Within a Regional System of Innovation. A Case Study in a Dutch Region. In *Learning Regions: Theory, Policy, and Practice*; Boekema, F.; Morgan, K.; Bakkers, S.; Rutten, R. Eds.; Edward Elgar: Cheltenham, 2000.

Moulaert, F.; Sekia, F. Territorial Innovation Models: A Critical Survey. *Reg. Stud.* **2003,** *37*, 289–302.

Murray, R.; Caulier-Grice, J.; Mulgan, G. *Open Book of Social Innovation*; The Young Foundation (NESTA): London, 2010. https://youngfoundation.org/wp-content/uploads/2012/10/The-Open-Book-of-Social-Innovationg.pdf

Ohmae, K. The Global Logic of Strategic Alliances. *Harv. Bus. Rev.* **1989,** *67* (2), 143–154.

Petrişor, I. I. *The Strategic Management. The Potentiological Approach* [Originally in Romanian as: *Management Strategic. Abordare potenţiologică*]; Brumar Publishing House: Timişoara, Romania, 2007.

Piore, M. J.; Sabel, C. F. *The Second Industrial Divide*; Basic Books: New York, 1984.

Porter, M. E. *The Competitive Advantage of Nations*; The Free Press: New York, 1990.
Porter, M. E. Clusters and the New Economics of Competition. *Harv. Bus. Rev.* **1998,** *76*, 77–90.
Porter, M. E. Location, Competition, and Economic Development: Local Clusters in a Global Economy. *Econ. Dev. Q.* **2000,** *14*, 15–34.
Porter, M. *On Competition;* Harvard Business Review Press: Brighton, 2008.
Prahalad, C. K.; Hamel, G. The Core Competence of Corporation. *Harv. Bus. Rev.* **1990,** *68* (3), 79-91. https://link.springer.com/chapter/10.1007/3-540-30763-X_14)
Putz, M. V. Strategic Cube of the Organization Competitive Advantage [Originally in Romanian: Cubul strategic al avantajului competitiv al organizațiilor]. MBA Thesis, Faculty of Economy Science and Business Administration, West University of Timişoara, 2017.
Putz, M. V. *The Strategic Cube of the Distinctive Advantage. Epistemological Approach*, Chapter 2 of the Present Monograph, 2019a.
Putz, M. V. *Strategic Innovating Paths for the Distinctive Advantage. The Changing Management faraway from Equilibrium*, Chapter 4 of the Present Monograph, 2019b.
Putz, M. V. *Global Strategies in the Knowledge Economy. The Case of R&D Sustainability in the European Union*, Chapter 10 of the Present Monograph, 2019c.
Putz, M. V. *Strategic Innovation in the Organization Governance. The 8-folding of the Mission Balance*, Chapter 9 of the Present Monograph, 2019d.
Putz, M. V. *The Strategic Cube of the Distinctive Advantage. Networks with Catastrophic Surfaces*, Chapter 3 of the Present Monograph, 2019e.
Robertson, R. Globalization or Glocalization? *J. Int. Commun.* **1994,** *1* (1), 33–52.
Sahin, F.; Robinson, E. P. Information Sharing and Coordination in Make-to-Order Supply Chain. *J. Oper. Manag.* **2005,** *23* (6), 579–598.
Sastry, M. A. Problems and Paradoxes in a Model of Punctuated Organizational Change. *Admin. Sci. Q.* **1997,** *42*, 237–275.
Schneider, S. L. Framing and Conflict. Aspiration Level Contingency, the Status Quo, and Current Theories of Risk Choice. *J. Exp. Psyhol. Learn. Memory Cognit.* **1992,** *19*, 1040–1057.
Scott, A. J. Production System Dynamics and Metroplolitan Development. *Ann. Am. Geogr.* **1982,** *72* (2), 185–200.
Sterman, J. D. Expectations Formation in Behavioural Simulation Models. *Behav. Sci.* **1987,** *32*, 190–211.
Sterman, J. D. *Business Dynamics. Systems Thinking and Modeling for a Complex World*; McGraw-Hill: New York, 2000.
Tatikonda, M. V.; Stock, G. N. Product Technology Transfer in the Upstream Supply Chain. *J Prod. Innov. Manag.* **2003,** *20*, 444–467.
Wernerfelt, B. A Resource-Based View of the Firm. *Strateg. Manag. J.* **1993,** *5* (2), 171–180.

CHAPTER 9

Strategic Innovation in the Organization Governance: The Eight-Folding of the Mission Balance

ABSTRACT

The sustainable and regenerative governance is presented in relation with the mission account, by relationship and partnership with the multiple stakeholders, in the postmodern strategic management, so promoting the inclusions and the dialog/with total transparency in the social act and the economic action. The qualitative aspects are presented at cognitive level, with an approach of the methodological quantification by introducing the so-called 8 × 8 mission matrix. It considers the network and dialog with the stakeholders quantifiable by binary correlations (yes/no or 1/0), yet with the informational energy evaluation (of relationship) stored, by both quantum and logistic modeling. This way, one identifies three different fundamental types of relationship (good, detrimental, and synergic). The connection with the cubic strategic potential in the distinctive advantage paradigm of the organizations is realized by applying the maximum path—minimal entropy principle, according to the present econo-ecological governance ethical message.

Motto:

"On one side we have the world in our pocket,
on the other side the world is spying on us!"

—Jean-Michel Jarre (2016).
The Story Behind "Exit" from
Electronica 2—The Heart of Noise.

9.1 INTRODUCTION

The postmodern economy is increasingly focusing on so-called *corporate citizenship* (Andriof and McIntosh, 2001). It includes social responsibility of business through an entire set of programs that an organization or an enterprise carries out in a given environmental, social, political, and legal context. It should provide public goods and services in harmony with the competitive advantage while having the value of the investments specific to the economic entity in question, so promoting and practicing ethical (trustworthy) governance for stakeholders and society (Crescenzi, 2002a).

Thus, from the investors' perspective (employers) and business developers, the *corporate citizenship* motivation (CC) "decomposes" on the *sustainable components at the organization level* (CERFE, 2001):

- Altruism (about 47.6%), through which the common good prevails, is profit by any means;
- Opportunity (approximately 22.2%), by which CC provides a sustainable business development environment;
- Development of the business space (about 30.2%), which is based on CC strengthens the core of the organization's productive activity, *Potential Theory* after Petrişor (2007).

On the other hand, from the stakeholders' perspective (shareholders, contractors, physical persons, and entities in legal relations with the organization or with a specific business/project, clients, local community, etc.), the development and implementation of the concept of *corporate citizenship* appears to be beneficial for any questionnaire surveys. It nevertheless follows the sustainable societal values (CERFE, 2001):

- *Responsibility for investment territory (~90.5%)*;
- *The sensitivity of the actors involved (~50.8%)*;
- *Advertising and promotion of business/project (~33%)*;
- *The mix of the above values (~71.4%).*

From the last idea asserted, we can see the disorientation of the postmodern society, possibly with the perception of *plutocracy* (as a form of government in which the state power is concentrated in the hands of the richest). One can observe nowadays the working syllogism (Marchi, 2002): *ideologies need politics, politics needs economy, economy needs marketing;* the *result is in mass exclusivity*! None of them needs reflection! Thus, the obvious need for counter-marketing, ecological postmodernism (in the sense of moral breathing, too)—counterbalancing the global economy (with total connection, but also prevailing on the total control, see the present Motto), by the means of the proximity or the local economy, shortening the distance from producer to consumer. From this point on, the economic agglomerations with a beneficial vision at the level of the *industrial district* makes sense. Such a construction is, in principle, based on cultural, technological, and innovation values; however, this is not necessarily the case for the *cluster or industrial park* concept—since they are promoters of cooperation and competition, the latter, being not always loyal, ethical, and therefore, sustainable economical behaviors.

In this context, a paradigm shift through *the nonprofit economy* of sustainable capitalism implies (Crescenzi, 2002b):

- *minimum entropy*;
- searching *for sense and meaning*;
- *diffuse environmental sensitivity*;
- *inclusion and integration*;
- culture of entrepreneurship and *entrepreneurship of the environment*;
- *incisiveness* through Porter's forces and competitive anti-forces (2008)—see the Hamming–Putz model (Putz, 2019a);
- *professionalism and competence*—through meritocracy, as a result of democracy, bureaucratically certified (see also these conclusions);
- *global opportunities* through revised marketing mix;
- *value,* as a size and social measure, *renewable* amount.

On the other hand, the *governance of social responsibility* is manifested through (Crescenzi, 2002c):

- strong identity, based *on the mission of the organization*;
- inclusion of public and private actors through *the culture of relationship;*
- overcoming idiocentrism by *allocentrism*: the manifestation of interdependence, sociability, family integrity, considering the needs and wishes of members of the belonging group, with the closeness in relations with the other members of the group;

- investing in "social capital" through research–development–innovation;
- highlighting (enhancing the potential) of the "organizational machine" (Petrişor, 2007);
- managerial training oriented toward social transformation.

Thus, it appears the possibility of going from *strategic governance to ethical (wise and postmodern) citizenship* (McCarney, 1996; Putnam et al., 1993; The Commission of European Communities, 2001):

- *Transparent approach* (open system): integration of all actors;
- *Participatory approach:* knowledge management—including networking;
- *Responsibility (accountability)*: cautious technical up-grading, e-inclusion, and e-government;
- *Efficiency*: global–local approach (GLocal);
- *Coherence*: capitalization of social and environmental capital.

Finally, the need to go through these stages of ethical governance requires specific strategic guidelines (Crescenzi, 2002c) as follows:

- Strategic *planning*
- *Design* of actions/activities
- The *accumulation* of social money (fund raising, Fig. 9.1);
- *Communication* and promotion campaigning
- Human resource *development* (people raising)
- *Recording* of economic activities as a tool of social legality
- *Partnership*

The present essay is, therefore, a plea for *gnoseocracy* (Fig. 9.2), as a form of organizing the business, and of its dynamic development in the business space. The strategic integration is two folded, namely, employing the research–development of innovation in the microeconomy (at the roots of the sustainable level) as well as on the macroeconomic level (at the integrated sustainable level). In this context, the knowledge-based decisions would work for harmonization between corporate and nonprofit actors, private and public, in an ethical (with an *ethos*), eclectic and lucrative governance. On the other hand, it is always the most difficult to know-how (how I can know?) and to understand–what (what I can do?)! Perhaps the present chapter may unveil some hints, in this regard.

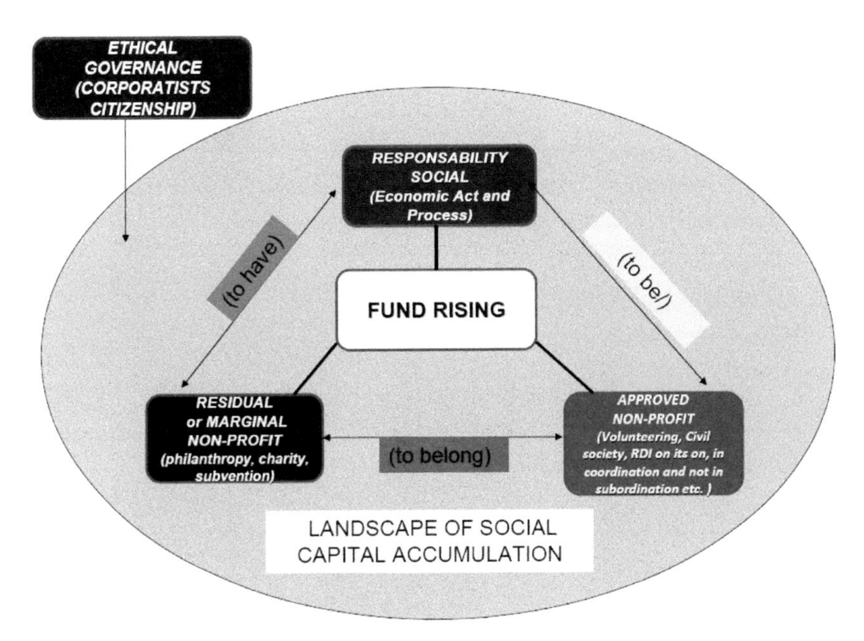

FIGURE 9.1 The triangle of the accumulation of social capital ("fund rising").
Source: Adapted with permission from Crescenzi (2002b).

FIGURE 9.2 Gnoseocratic area of governance–residual–homologation correlations in the context of confronting associated paradigms.
Source: Adapted with permission from Crescenzi (2002c) and Hina (2002).

9.2 COGNITIVE ANALYSIS

It starts from the *triptych of ethical value*:

Vision (dream) → Mission/project → Spatio-temporal solidarity (by locating/relocating/relocating the business)

And it is further individualized:
Vision (dream) with the components:

- simple: for a specific problem—a specific solution
- complex: more services (economic iteration) for more usages (temporal iteration) in several places (spatial iteration)

and the *mission* (*volere–dovere–potere, in Italian a.k.a. "to whish to—to have to—to can to" in English*) which introduces the mission balance sheet, possibly in relation to the potentiality theory of organizational strategies (Petrişor, 2007), Figure 9.3:

- wanting: strategy
- having to: organize
- to can versus could/might: active potential (internal organization) and latent (in relation to the environment)

With regard to solidarity with local, regional, and national communities "a mouse of solidarity" [*A topolino di soliedarità*, originally in Italian], (Rey, 2002), can be realized through the so-called third sector. It should activate the nonprofit component of sustainable organizations seen as the organization actions and economical behavior oriented to live better with stakeholders (and not necessarily toward profit oriented by any means; Fig. 9.4).

In this case of economic solidarity with the stakeholders, one can consider the social balance sheet related to the mission balance, for which the stages of strategic development of the business/economic behavior are identified as follows (Fig. 9.5):

- *passive strategy*: the interests of the organization
- *reactive strategy*: relationship with stakeholders
- *iterative strategy*: partnership with stakeholders
- *proactive strategy*: sustainable development.

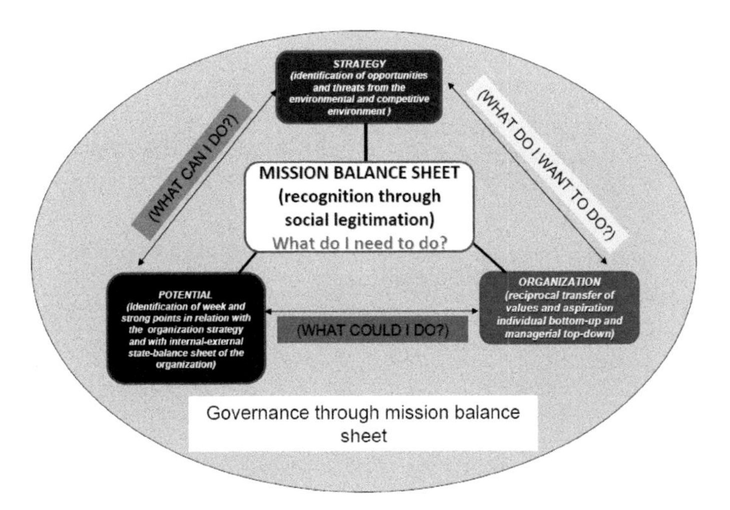

FIGURE 9.3 Triptych of the mission balance sheet.

Source: Inspired by Crescenzi (2002d), Hina (2002), and Petrisor (2007).

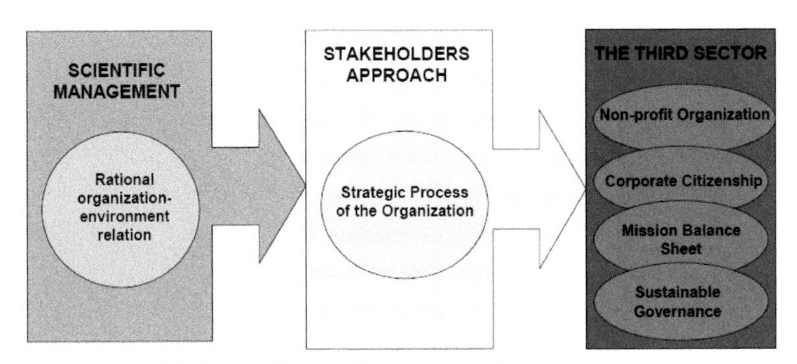

FIGURE 9.4 The third sector (nonprofit management).

Once the "nonprofit" paradigm is accepted, it is possible to move from the mission balance to mission governance, designed on the same potentials' triptych as in Figures 9.1 and 9.3 (Fig. 9.6):

- *the ethical values*: declared values, patrimony management, organizational primary processes… the question *how*? is answered
- *the facts*: answer the question *what*? In relation to the purpose and mission stated in the strategy
- *the figures*: answer the question *with what results*? For all organizational/institutional activities

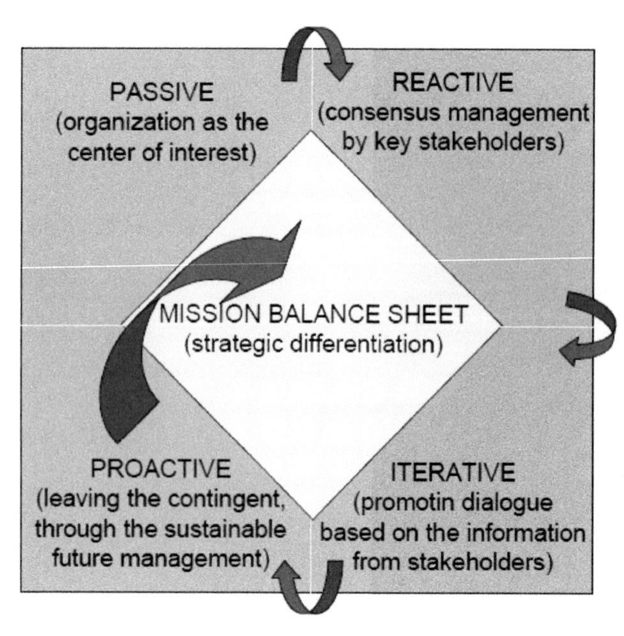

FIGURE 9.5 Mission balance: strategic differentiation (see also Chapter 8 of the present monograph).

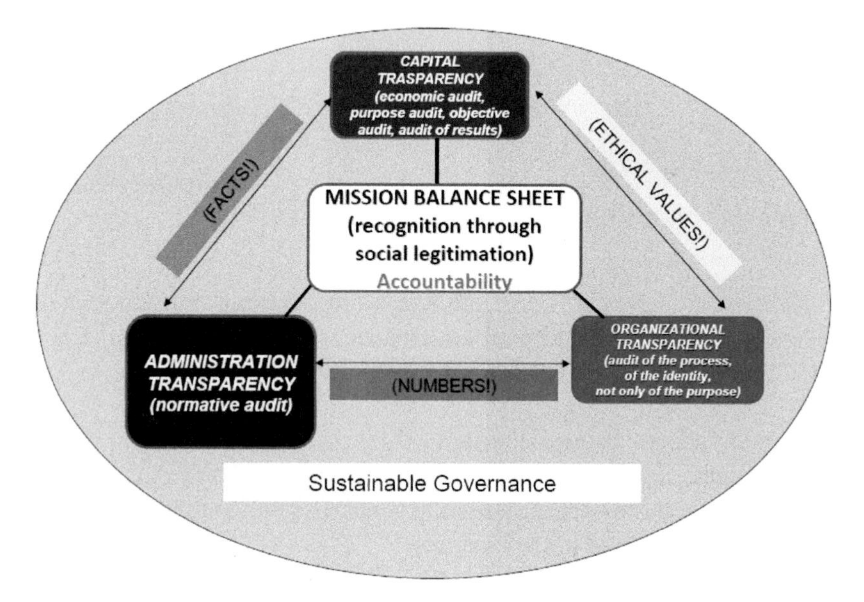

FIGURE 9.6 Mission balance: differentiation of governance.

Finally, the triptych of ethical action (Fig. 9.7) is reached:

Mission (what) → Governance (how) → Balance sheet (with what results)

which allows the subsequent conceptual–quantitative approach (Section 9.3) for the organization's "implementation" of governance.

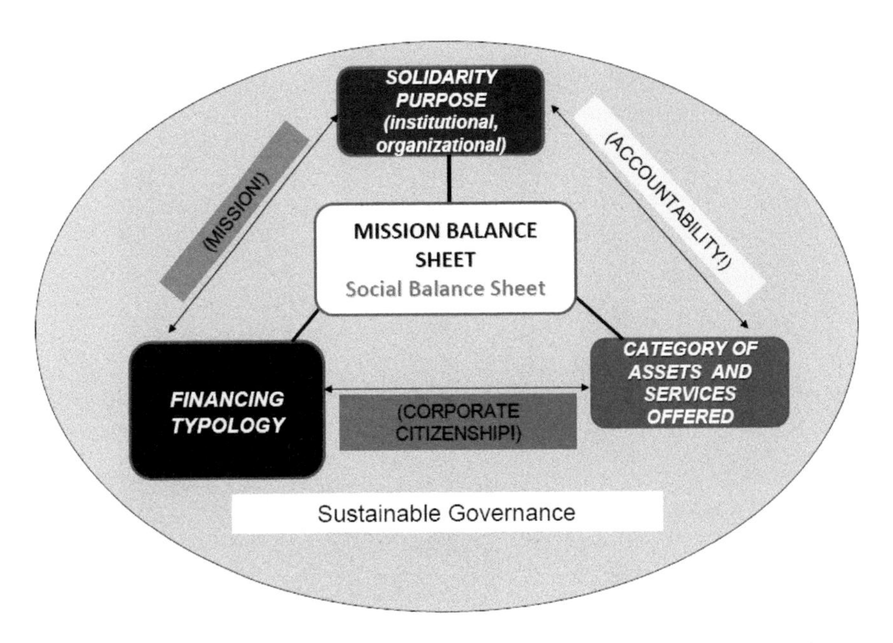

FIGURE 9.7 Mission balance sheet: the social balance of the organization/enterprise.

9.3 METHODOLOGICAL ANALYSIS

It is based on the Copenhagen charter theory of reporting to the stakeholders in eight steps of Figure 9.8, as adapted upon The Copenhagen Charter (1999):

1. *approval of top management*
2. *identifying key stakeholders and critical factors of success and value*
3. *dialogue with stakeholder*

4. *determining* the key success indicators and adapting the information management system
5. monitoring organization performance and stakeholders' satisfaction based on structural values (and activated potential)
6. formulation of objectives, budget, and action plans to improve the previous step
7. preparation, verification, and publication of the organization's balance sheet (auditing)
8. Consult stakeholder on performance, values and targets for performance improvement
9. *Resuming the stakeholder relationship cycle.*

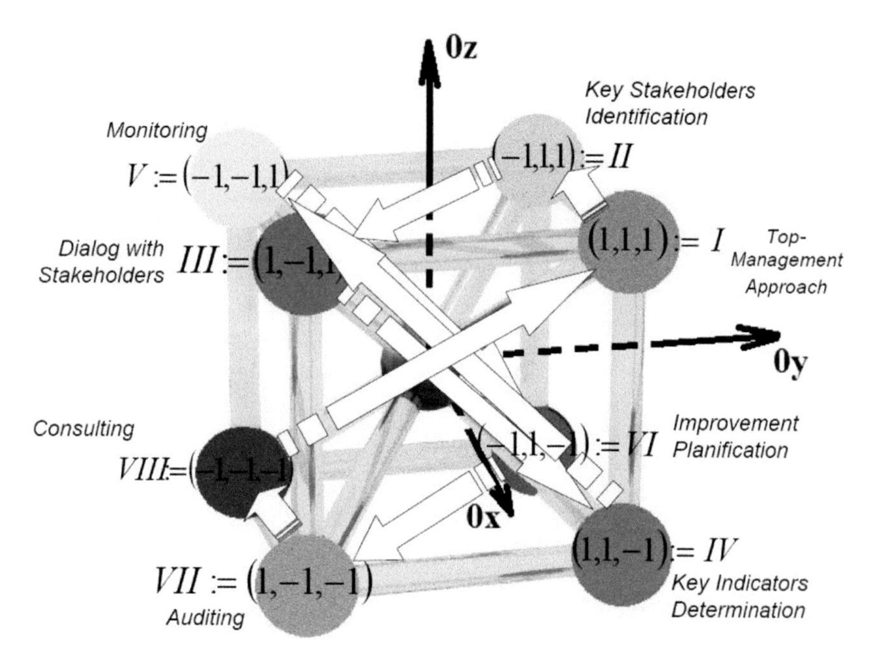

FIGURE 9.8 The Copenhagen Charter (1999) of governance in relation to stakeholders enlisted in the distinctive advantage cube.

Source: Adapted with permission from Putz (2017, 2019b–2019c).

Mission Balance Governance, in its present approach, is in line with the eight informational stages of the Copenhagen Charter (Hina, 2002):

1. the action plan
2. network of stakeholders
3. the maximum path (minimum entropy), see the "path" illustrated in Figure 9.8
4. the mission report
5. the mission balance sheet
6. social audit
7. improvement plan
8. publication of results.

Based on this information, the mission balance matrix can be entered in relation to stakeholders or shortly called *the mission matrix*, Table 9.1.

In terms of binary matrix, Table 9.1 is synthetically rewritten as

$$A_{MB} = \begin{pmatrix} 1 & 1 & 0 & 0 & 0 & 0 & 0 & 0 \\ ? & ? & ? & ? & ? & ? & ? & ? \\ ? & ? & ? & ? & ? & ? & ? & ? \\ 0 & 0 & 1 & 0 & 0 & 1 & 1 & 0 \\ 0 & 0 & 1 & 1 & 1 & 1 & 1 & 0 \\ 0 & 0 & 0 & 0 & 0 & 0 & 0 & 1 \\ 0 & 0 & 0 & 0 & 0 & 1 & 0 & 1 \\ 0 & 0 & 0 & 0 & 0 & 0 & 1 & 0 \end{pmatrix} \tag{9.1}$$

with the value of mission energy calculated in quantum and logistic manner and conceptually exemplified here in Table 9.2.

We rank the results according to the minimum occupancy (a.k.a. minimum entropy) of the correlations (here in the binary logic "0" or "1") of the stakeholder network and their maximum paths in relation to the entries in the Copenhagen Charter.

Subsequent studies can explore other possibilities of arranging–pairing–linking the stakeholder network with the maximal path of information top management-driven process, and appropriate interpretation in light of the examples presented here.

9.4 CONCLUSIONS AND PERSPECTIVES

Happiness can be never global! Unfortunately! Instead, advertising is a value of use, "communication that lies by definition"; developed by

TABLE 9.1 Mission Matrix with "Degrees of Freedom" in the Entries of Stakeholder's Network, and the Maximal Path Between Them.

	Top-management Approval	Identification key stakeholders	Dialogue with stakeholders	Determining key indicators	Monitoring	Improvement planning	Auditing	Consulting
Action plan	1	1	–	–	–	–	–	–
Stakeholders network	?	?	?	?	?	?	?	?
Maximal path	?	?	?	?	?	?	?	?
Mission report	–	–	1	–	–	1	1	–
Mission balance	–	–	1	1	1	1	1	–
Social audit	–	–	–	–	–	–	–	1
Improvement plan	–	–	–	–	–	1	–	1
Publishing of results	–	–	–	–	–	–	1	–

TABLE 9.2 Synopsis of Quantum and Logistic Models for the Mission Energy Associated with the Mission Matrix (8 × 8) and Its Determinant, Reflecting the Degree (Potential) of the Development of the Act of Governance in Relation to the Network of Stakeholders in Its Dynamics and Its Inertia to the Variables of Copenhagen Governance.

Mission balance matrix	Mission determiner	Mission energetic model	
		Quantum	**Logistic**
		$I_{MB}=\left(\det[A_{MB}]\right)e^{-\det[A_{MB}]}$	$I_{MB}=\frac{1}{\det[A_{MB}]+e^{-\det[A_{MB}]}}$
		Interpretation of results	
$A_{MB}=\begin{pmatrix}1&1&0&0&0&0&0&0\\0&0&1&0&1&0&0&0\\1&0&0&0&1&0&0&0\\0&0&1&0&0&1&1&0\\0&0&1&1&1&1&1&0\\0&0&0&0&0&0&0&1\\0&0&0&0&0&1&0&1\\0&0&0&0&0&0&1&0\end{pmatrix}$	$\det[A_{MB}]=1$	$I_{MB}=0.368$	$I_{MB}=0.731$
		The positive determinant is *beneficial* (kinetic catalyst) for stakeholder dialogue and monitoring, subject to the approval of top-management monitoring of the maximum intra-network pathway! The mission energy proves to be high, but not maximum	
$A_{MB}=\begin{pmatrix}1&1&0&0&0&0&0&0\\0&0&1&0&0&1&0&0\\1&0&0&0&1&0&0&0\\0&0&1&0&0&1&1&0\\0&0&1&1&1&1&1&0\\0&0&0&0&0&0&0&1\\0&0&0&0&0&1&0&1\\0&0&0&0&0&0&1&0\end{pmatrix}$	$\det[A_{MB}]=0$	$I_{MB}=0$	$I_{MB}=1$
Considering the matrix with only one unit moved from monitoring to the right in the improvement plan in correlating the stakeholder network with Copenhagen inputs		The mission energy is maximal (synergic) by modifying the stakeholder network to improve it, with "zero" inertia, detrimental to improvement, while the maximum track is monitored and approved by top-management!	

TABLE 9.2 *(Continued)*

Mission balance matrix	Mission determiner	Mission energetic model	
		Quantum	**Logistic**
		$I_{MB} = (\det[A_{MB}])e^{-\det[A_{MB}]}$	$I_{MB} = \frac{1}{\det[A_{MB}] + e^{-\det[A_{MB}]}}$
		Interpretation of results	
$A_{MB} = \begin{pmatrix} 1 & 1 & 0 & 0 & 0 & 0 & 0 & 0 \\ 0 & 0 & 1 & 1 & 0 & 0 & 0 & 0 \\ 1 & 0 & 0 & 0 & 1 & 0 & 0 & 0 \\ 0 & 0 & 1 & 0 & 0 & 1 & 1 & 0 \\ 0 & 0 & 1 & 1 & 1 & 1 & 1 & 0 \\ 0 & 0 & 0 & 0 & 0 & 0 & 0 & 1 \\ 0 & 0 & 0 & 0 & 0 & 1 & 0 & 1 \\ 0 & 0 & 0 & 0 & 0 & 0 & 1 & 0 \end{pmatrix}$ Considering the matrix with only one unit moved from monitoring to the left in determining the key indicators in correlating the stakeholder network with Copenhagen inputs	$\det[A_{MB}] = -1$	$I_{MB} = 2.718$ The mission's energy is *detrimental* for the success of the organization and for its economic process; the organization apparently abandoned monitoring the progress toward determining key indicators for the stakeholder network. The negative determinant indicates an increase in process inertia and a decrease in the success of governance through the mission balance sheet. At the same time it also indicates an over-stabilization of the stakeholder network which, in social terms, means that they are opaque to social transparency and hence to the quality of governance itself!	$I_{MB} = 0.582$

so-called *fast-thinkers* (who have an answer to anything, especially when they cannot give it) based on the trading of information, both as news and as merchandise! Thus, the postmodern society evolves within the ideological triptych:

- *Bureaucracy*: based on the principle (need) of certainty ... past time
- *Democracy*: based on the precautionary principle (need) ... present time
- *Meritocracy*: based on the principle (reality) of uncertainty ... future time

resulting in a risk perceived as increased when moving from the future to the present—threatening to destroy the past (consolidated structures, and organizations characterized by inertia to change). Paradoxically, it is precisely the formula of *social risk perceived* as being coined by Sandman (2012). The "contingent hazard" ("the hazard" includes environmental scientific measures) increases the "public anxiety" ("the outrage," panic, anger, horror, and uncontrollable reactions of society). The strategic management including the balance mission is providing the solution to the improvement of the effects of uncertainty: the social inclusion by ethical cooperation, governance through corporate citizenship. This study approached these issues in both qualitative (in eighth dimensional space) and quantitative (by algebraic matrix computations) manner.

Here is how designing the eight-step impact of an innovative governance act based on the strategic mission (mission balance) is feasible—and implies, by generalizing nonprofit management (Crescenzi, 2002e):

- *coherence between the project and the mission:* the correspondence between the internal and external context of the business and organization, the *sustainability* of the organization as level-binder between the competitive advantage and the regenerative advantage (Putz, 2017)
- *sustainability:* impact on society
- *bottom-up approach:* through participatory management
- *Key partnership: at horizontal (national) and vertical (international and European) level*
- *critical resources:* good practices
- *competitive advantage:* technology (satellite connection) or value (through ethical behavior and trust generator)

- *geographic innovation* (stakeholder network and clustering): structural (reorganization of the firm) and strategic (new management tools—report and mission balance sheet type)
- *evaluation of the project:* by complex indexes, in the space of the mission balance sheet; see the mission matrix, the determinant and the energy of the mission in eq 9.1 and Tables 9.1 and 9.2.

Thus, *ethical governance*, based on the mission balance, in the context of Euro-design, necessarily meets the following metaphors (Risso, 2002):

- *The concert*—the scene and the context are accommodated in *eight steps:* from (1) idea (the social dream and the economical vision); to (2) the reliability (the available knowledge, skills, and motivations); by the aid of (3) the team (the achievement of individual group is inclusively important, such as the motto of the 1995 international year of tolerance: "we are all different, therefore equal!" in the great econo-ecologic life cycle); by developing (4) the project (while assuming specific roles, obligations, timing and development tools for the "melody" a.k.a. development—assuring in this way, the flowing-running of the project by integrating the "samples" into "rehearsals"); assuring the (5) project coordination (by the postmodern manager—the postmodern orchestra conductor, that is, the "project director," in fact the true guarantor of harmony between different instruments, methods and techniques of management); in a limited, yet not a limitative (6) budget (with its subdivisions as "general rehearsal"); while communicating the result through (7) the presentation of the project (in front of private or public stakeholders, such as an "preview"); toward the next step, that is, (8) the project selection—stakeholder decision to finance (this way assumed as a group success).
- *The water walking*—exercises the alternative strategic management, so-called strategic intent, respectively, managed in eight stages: (1) accepting uncertainty, the risk and challenge, (2) managing with self-control the competence and the talent; (3) moving and combining various cultural, ethical (trusted), and political frameworks; (4) creating contacts, networks, and interventions; (5) keeping the mission to go through a programmatic roadmap; (6) continuously reinventing the strategic stages, (7) firmly pointing on flexible points that may enhance the application methodology

(tactics), and (8) integrating the available time space and comprehensive means to achieve and maintain the successful results.

With these innovative metaphors, the dynamic context of organizing a business project, enterprise, and even agglomerations of economic entities (industrial districts or clusters, all involving cutting technology or continuously research–development–innovation) is no longer an "adventure," but rather a process integrating the methodology, cultures, and ethics such that they all become inherent components of compatibility—compensation of the developmental mission for the wise society (triple characterized by efficient communications, renewable energy, and a long and healthy life). Therefore, the strategic managerial vision should come close to the scientific innovation by integrated research of the internal context with the external environment of the organization/firm/business; only by this way, one can envisage an "economic life with environmental friendly viability, along the social recognition and acceptability." The praise for the ethical projection of business governance through wise management of the competitive, sustainable, and regenerative advantage is, thus, perennially challenged.

KEYWORDS

- **governance**
- **strategic mission**
- **the third sector**
- **sustainability and regeneration**
- **informational energy**
- **strategic cube of the distinctive advantage**

REFERENCES

Andriof, J.; McIntosh, M. *Perspectives on Corporate Citizenship*; Greenleaf Publishing: Sheffield, UK, 2001.

CERFE. Action-Research on Training Needs and Best Practices on Corporate Citizenship Among European Small and Medium Enterprises. Gruppo CERFE (Centro di Ricerca e Documentazione Febbraio'74): Roma, 2001. http://www.cerfe.org/public/CorCit1.pdf.

Crescenzi, M. Le nuove frontiere della corporate citizenship. In *Manager & Management Non Profit—La Sfida Etica*; Crescenzi, M., ed. (ASVI—Agenzia per lo Sviluppo del Non Profit—Onlus, www.asvi.it); ASVI Editoria: Roma, 2002a.

Crescenzi, M. Un viaggio del non profit. Prologo In *Manager & Management Non Profit—La Sfida Etica*; Crescenzi, M., Ed. (ASVI—Agenzia per lo Sviluppo del Non Profit—Onlus, www.asvi.it); ASVI Editoria: Roma, 2002b.

Crescenzi, M. Il paradigma della governance. Per un nuovo modello d'azione delle organizzazioni non profit italiane, tra superamento della residualità e rifiuto dell'omologazione. In *Manager & Management Non Profit—La Sfida Etica*; Crescenzi, M., ed. (ASVI—Agenzia per lo Sviluppo del Non Profit—Onlus, www.asvi.it); ASVI Editoria: Roma, 2002c.

Crescenzi, M. Coltivare sogni e raccogliere impatti. Vision, mission, progetti e risultati coerenti. In *Manager & Management Non Profit—La Sfida Etica*; Crescenzi, M., Ed. (ASVI—Agenzia per lo Sviluppo del Non Profit—Onlus, www.asvi.it); ASVI Editoria: Roma, 2002d.

Hinna, A. Il bilancio di missione da strumento di misurazione a leva di governo dell'organizzazione. In *Manager & Management Non Profit—La Sfida Etica*; Crescenzi, M., Ed. (ASVI—Agenzia per lo Sviluppo del Non Profit—Onlus, www.asvi.it), ASVI Editoria: Roma, 2002.

Marchi, A. Per una "competiţione etica": dal controllo sul prodotto alla gestione del processo. In *Manager & Management Non Profit—La Sfida Etica*; Crescenzi, M., Ed. (ASVI—Agenzia per lo Sviluppo del Non Profit—Onlus, www.asvi.it); ASVI Editoria: Roma, 2002.

McCarney, P. L. *Changing Nature of Local Governement in Developing Countries*; Centre for Urban and Community Studies, University of Toronto and the Federation of Canadian Municipalities International Office: Toronto, 1993.

Petrişor, I. I. *The Strategic Management. The Potentiological Approach* [Originally in Romanian as: *Management Strategic. Abordare potenţiologică*]; Brumar Publishing House: Timişoara, Romania, 2007.

Porter, M. *On Competition;* Harvard Business Review Press: Brighton, 2008.

Putnam, R.; Leonardi, R.; Nanetti, R. Y. *Making Democracy Work: Civic Traditions in Modern Italy*; Princeton University Press: Princeton NJ, 1993.

Putz, M. V. *Strategic Cube of the Organization Competitive Advantage* [Originally in Romanian: *Cubul strategic al avantajului competitiv al organizaţiilor*]. MBA Thesis, Faculty of Economy Science and Business Administration; West University of Timişoara, 2017.

Putz, M. V. *The Strategic Cube of the Distinctive Advantage. Epistemological Approach*, Chapter 2 of the Present Monograph, 2019a.

Putz, M. V. *The Strategic Cube of the Distinctive Advantage. Networks with Catastrophic Surfaces*, Chapter 3 of the Present Monograph, 2019b.

Putz, M. V. *Strategic Innovating Paths for the Distinctive Advantage. The Changing Management faraway from Equilibrium*, Chapter 4 of the Present Monograph, 2019c.

Rey, R. R. *Non profit?* iO-Donna, 26 October (in Italy), 2002.

Sandman, P. M. *Responding to Community Outrage: Strategies for Effective Risk Communication* First published in 1993 by the American Industrial Hygiene Association.;

Copyright transferred to the author, Peter M. Sandman, in 2012. The full text of this book is online at: http://psandman.com/media/RespondingtoCommunityOutrage.pdf.

The Commission of European Communities. *The European Governance—A White Book*; European Commission: Bruxelles. http://europa.eu/rapid/press-release_DOC-01-10_en.htm, 2001.

The Copenhagen Charter launched at *Building Stakeholder Relations—the Third International Conference On Social and Ethical Accounting, Auditing and Reporting* Taking Place in Copenhagen, Denmark, on November 14–16, 1999. The Conference Was Co-hosted and Organized by the INSTITUTE of Social and Ethical AccountAbility, Novo Nordisk A/S, The Copenhagen Centre, Copenhagen Business School and the House of Mandag Morgen, 1999, http://www.improntaetica.org/file/docs/copenhagencharter.pdf.

CHAPTER 10

Global Strategies in the Knowledge Economy: The Case of R&D Sustainability in the European Union

ABSTRACT

In the total globalization context, the main globalization factors are highly present. The global competitiveness index (GCI), the productivity index in the sociopolitical context (KOF), the world investment report index (WIR), just to select the most preeminent in the globalization quantification, are correlated with the research and developing index (R&D) as represented in the gross domestic product (GDP) percent of a country—eventually in a glocal network (with the typical example of the European Union). The last one may be seen as a dynamic effector among the firsts (as the first instance effect) and with an enhancer (amplification effect), in a further phase. The correlations between them are numerically explored, by means of the multilinear regression formalism, with implicit orthogonality. Thus, the so-called quantitative knowledge–globalization relationship is established. The correlation "spectrum" (i.e., the uni-, bi-, and trilinear unfolds in this study) allows the spectral-KGR approach, by which the minimal path (as optimal strategy) of the directions of the selected hierarchy is selected (especially the "first wave"—as alpha strategic moving—for the direct causality in knowledge—globalization). Nevertheless, the second and the third hierarchies, that is, the "beta and gamma" strategic waves, are considered as co-lateral and supporting effects; at the end, the optimal hierarchy competitiveness → investments → productivity is projected into the glocal picture, in a firm-developing strategy (and/or of a business) with the identification of the minimum path in the strategic cube of the distinctive advantage. The relation with the maximum path—minimum entropy principle is also observed, and the perspective of the min–max unification in a theory of the postmodern knowledge economy, always in a transition state, of proximity, or as a nexus search, is also suggested.

Motto:

"Why this? Why that? And why?"

—Jean-Michel Jarre (2016).
—*Electronica 2—The Heart of Noise.*

10.1 INTRODUCTION

What variables are essential to substantiate a modern (post) democracy theory? And what role does globalization have in this relationship if it exists?

The (interim) answer to these questions can be formulated by testing the extreme bounds of democracy in the globalized society (Gassebner et al., 2012):

- *Transition to democracy* (Przeworki et al., 2000);
- *The survival of democracy* (Przeworski, 2005).

These limits depart from—and are embraced by—the main theories of the transition to democracy:

- *Cultural* approach (Almond and Verba, 1963);
- *Economic* approach (Lipset, 1959) and theorizing "modernity" (Huntington, 1968);
- *The ecological* approach through the strategic exploitation of natural resources (Boix, 2003);
- Approach through *the theory of political, cultural, and economic diffusion* (Gleditsch, 2002).

Within the diffusion theory, the analysis of extreme limits of democracy through first-order Markov processes uses the conditioned probability of democracy at time t ($D = 0$ if does not exist and $D = 1$ if it exists):

$$\Pr\left(D_{i,t}\left|D_{i,t-1}\right.\right)=\left(1-D_{i,t-1}\right)\Pr\left(D_{i,t}\left|D_{i,t-1}=0\right.\right)+D_{i,t-1}\Pr\left(D_{i,t}\left|D_{i,t-1}=1\right.\right). \quad (10.1)$$

By developing the probability amplitudes in a multiline manner (in the generic variable "x" designating the commonly accepted explanations of the evolving democratic phenomenon), the transition from authoritarianism to democracy (AD) and the survival of democracy-to-democracy (DD) is shaped:

$$\begin{cases} \Pr\left(D_{i,t} \middle| D_{i,t-1} = 0\right) = \sum_{x} \beta_x^{AD} x_{t-1} \\ \Pr\left(D_{i,t} \middle| D_{i,t-1} = 1\right) = \sum_{x} \beta_x^{DD} x_{t-1} \end{cases} \quad (10.2)$$

Based on a database of 165 countries, over a timeframe between 1976 and 2002, analytical results are obtained, that support the essential aspects of the impact of globalization on the dynamics of democracy (Gassebner et al., 2012):

- For emerging democracies, GDP per capita does not explain democratic transitions. In these cases, per capita GDP is not even a "border" variable, which suggests that prosperity is not automatically associated with democracy, as illustrated by the Arab revolutions
- An earlier democratic experience increases the likelihood that democracy will emerge again (e.g., OECD countries, which, as an organization, demand from the start a democratic experience)
- Economically performing countries are less inclined for democratic emergence (in contrast to what modern democracy advocates in democracy), the trend being to promote autocracy
- For democratic countries, instead, GDP and economic development motivate the survival of democracy
- Neighborhood with democratic countries can be a catalyst for the emergence and/or survival of democracy.

The overall conclusion on democracy suggests that short-term economic development is not a solution, instead in the long run, economic growth contributes significantly to the survival of democracy.

More generally, the determinants of sustainable growth in a nation, region, economy (and possibly clusters) are identified as (Dreher, 2006a):

i) *advantage of proximity through exchanges* (economic, social, and political)
ii) *the flow of capital* and/or the opening to these flows

with a role in diminishing the *unfairness of globalization* (Heineman, 2000) and finally in the *empowerment* of democracy (Petrişor, 2007).

Estimating "cross-sections" for these factors involves correlating the economic–political–social activity in question with their individual effect on globalization (the glocal effect), avoiding *the issue of endogenity* (the issue of the reversed causality, that is, assuming cause as effect and/or the other way round.

Hence, the so-called index of the Swiss business cycle—*Konjunktur-forschungsstelle—KOF* of globalization (Dreher, 2006a, 2006b; Dreher and Gaston, 2008), which, in the current version, cumulates these influences (deterministic causes), according to the KOF-20163 diagram in Table 10.1.

Other globalization indices are introduced, such as:

- Globalization Index Kearney/Foreign Policy Magazine (Kearney, 2001)
- Globalization Index *CSGR—Center for the Study of Globalization and Regionalization* at the University of Warwick (Lockwood and Redoano, 2005)
- The cultural globalization index (Kluver and Fu, 2004)
- The Maastricht globalization index (Figge and Martens, 2014)
- The new globalization index (Vujakovic, 2010)
- The person-based globalization index (Caselli, 2012)
- Relational/Connected indices are rather *connectedness* than globalization, for example, UNIDO Connectedness Index (UNIDO, 2016), and DHL Global Connectedness Index (DHL 2014), the KOF index is mostly used, despite the fact that it does not take into account global impact variables, such as:
  - Migration of the population
  - Religion, especially Islamic globalization, historically motivated—at the 1500th anniversary of its founding—following the same cycle of reform and expansion, as Christian religion was in the Middle Ages, see Aron (1961)
  - Ecological footprint present in the Maastricht Index, including production bioavailability
  - Influence of knowledge through the globalization of growth–development–technical–scientific innovation

TABLE 10.1 Composition of the KOF-2013 Globalization Index, with the Subindices, Variables, and Subvariables Weighted by (%) Cumulative Percentage.

(%)	Globalization subindex	Variable		Subvariable	
		Name	(%)	Name	(%)
36%	Economic globalization	Actual flows	50%	Trade (percent of GDP)	22%
				Foreign direct investment, stocks (percent of GDP)	27%
				Portfolio investment (percent of GDP)	24%
				Income payments to foreign nationals (percent of GDP)	27%
		Restrictions	50%	Hidden import barriers	23%
				Mean tariff rate (on all available market products)	28%
				Taxes on international trade (percent of current revenue)	26%
				Capital account restrictions (restrictions on capital flux reflecting the national property assets, owners on selling—the positive income flux in country, and foreigner at buying—negative outcome flux from country)	23%
37%	Social globalization	Data on personal contact	33%	Telephone traffic	26%
				Banking transfers (percent of GDP)	2%
				International tourism	26%
				Foreign population, including immigration (percent of total population)	21%
				International letters (per capita)	25%
		Data on information flows	35%	Internet users (per 1000 people)	36%
				Television (per 1000 people)	38%
				Trade in newspapers (percent of GDP)	26%

TABLE 10.1 *(Continued)*

(%)	Globalization subindex	Variable		Subvariable	
		Name	(%)	Name	(%)
		Data on cultural proximity	32%	Number of McDonald's restaurants (per capita)	46%
				Number of IKEA stores (per capita)	46%
				Trade in books (percent of GDP)	8%
27%	Political globalization	Embassies in country	25%		
		Membership in international organizations	27%		
		Participation in UN Security Council Missions	22%		
		International treaties	26%		

Source: Adapted with permission from KOF Index of Globalization (2016).

- income inequality, see Stiglitz (2002)
- globalization of human rights
- globalization of gender equality
- globalization of poverty, and violence versus Tolerance (Berggren and Nilsson, 2015)
- globalization of the expected life horizon
- globalization of diseases
- globalization of terrorism
- globalization of competition
- globalization of financial crises (Chotikapanich et al., 2012).

However, *hyper-globalization* is not possible in absolute terms. This happens because encompassing everything in a single indicator—it no longer correlates with anything, becoming cause and effect, which in cognitive terms would self-annihilate. The concept would be meaningless and useless, while in the practical terms, it would unify everything, with no possibility to perceive the change (necessary for evolution and adaptation to it), and, therefore, no progress can be recorded! It would give a very ungodly, insensitive, unproductive, undiscernibly *status quo*—a form of *socialist capitalism,* or maybe even (through the reduction to absurd) of communist capitalist (nationalizations or global confiscations included)! It would be the climax of capitalism, perhaps Marx's dream unified with that of Lenin's, to reach communism through capitalism! But, it would also correspond, *mutatis mutandis,* to global overheating, which, after melting all planetary glaciers and levels of the temperature of the planetary ocean, opens the way for (the biggest) glaciations process, the total frost, having as a consequence the extinction of species! Likewise, communist capitalism (as an extreme form of hyper-globalization) will lead to the freezing of economies, poverty, inequities, the start of civil wars within and internations, to the final desert..., unfortunately the same Ice Age! It would be an unfortunate economic and ecological unification in the extinction of the human species, and that too long before the solar energy would be consumed!

Excluding the apocalyptic scenario of hyper-globalization, the KOF index must, and can be correlated with other globalization indices, highlight the degree of correlation (through the "collision areas"), between this and various other globalization indices in a cognitive aggregation. Thus, it can only be considered a strategic index (with a managerial sense of direction and facilitating development opportunities), and ultimately, also an economic index (by increasing efficiency in obtaining, circulation and re-cycling of goods and services)—in a durable (local) and sustainable (global) glocal sense, at the level of companies and nations. The present chapter is in line with R&D investments, its global causes–effects, and identifying business development strategies (firms, clusters, and multinational).

10.2 COGNITIVE ANALYSIS

Is globalization a beneficial or detrimental factor for well-being?

But for the national, local, regional, and global poverty?

The issues of measuring globalization (especially in relation to poverty, *globalization-poverty nexus*) are (Bergh and Nilsson, 2014):

- Measuring economic growth rather than poverty.
- The paradox of the endogenity of globalization (by manifesting both as cause and effect of income growth).

The J-inverted curve of globalization influence on poverty depends on *short-term factors* (Agénor, 2004):

- transition costs
- shortage of human capital
- inflation
- social costs: the hypothesis of minimum (minimal state)—*the bottom hypothesis* versus compensation mechanism—*the compensation hypothesis*.

However, in the long run (10–15 years), the correlation of poverty with the aggregation of globalization factors in the KOF + index of other globalization factors

$$Poverty = \alpha + \beta(KOF) + \sum_i \gamma_i GLOBAL_i \tag{10.3}$$

generates a negative trend, indicating a detrimental influence on poverty, so beneficial for economic, social, and cultural development (Bergh and Nilsson, 2014).

Furthermore, on the generational models, the globalization indices measure (Potrafke, 2015):

- *first generation*: actual correlations.
- *second generation*: develops causality tests, extreme links, and sophisticated econometric procedures.

Geographically, globalization investigates, in compliance with *UNCTAD—United Conference on Trade and Development,* the foreign direct investment (FDI) for three categories of countries, as reported in the *World Investment Report* (WIR, 2015), as it follows:

- *Developed economies*: OECD members (except Chile, Mexico, the Republic of Korea, and Turkey) + the new European Union non-OECD countries (Bulgaria, Croatia, Cyprus, Latvia, Lithuania, Malta, and Romania) + Andorra, Bermuda, Lichtenstein, Monaco, and San Marino.
- *Countries with economies in transition*: Southeast Europe, Commonwealth Independent States, + Georgia.
- *Developing economies*: all economies not included above, excluding Hong Kong SAR, Macao SAR, and Taiwan Province of China.

In this context, investment policies are based and measured, in particular, on investment policies focused on *liberalization, promotion, and facilitation* (as attributes specific to management, rather than to general economy). In fact, countries and regions are engaged in a comprehensive reform of the international investment agreements (IIA) in order to address five major global challenges:

- *"safeguarding"—ensuring the right to regulate the public interest* so that the influence of the IIA on the sovereignty of the countries concerned does not affect public policies.
- *Reforming the current dispute over the nature of the investments,* so as to explain the crisis of the current system.
- *Promoting and facilitating investments* to increase their volume.
- *Ensure responsible investments* so as to maximize the FDI effect and minimize potential adversity.
- *Increasing the systemic consistency of the IIA regime,* so as to ensure consistency in investment relations and the overall FDI system in general.

The consequence of these policies can lead to the regulation, control, and coherence of international tax treatment, especially in relation to the contribution of multinational corporations in developed countries, the leakage of tax avoidance through tax avoidance—possibly through *offshore* investments. Globally, in line with WIR-UNCTAD-2015, in 2014, there was a fall in FDI by 16% to $1.2 trillion; yet, it covered about 40% of the developed and transition countries' financial markets.

However, international sustainable development cannot be achieved in the absence of global competitiveness. In this regard, the *World Economic Forum*, "committed to improving the state of the world" reports the Global Competitiveness Index—GCI (2015). Like the KOF index, the GCI index contains three sub-indices, namely:

- GCI-1: *Basic requirements subindex*
- GCI-2: *Efficiency enhancers subindex*
- GCI-3: *Innovation and sophistication* factors subindex.

Each of these sub-indices is based on certain "pillars," which cumulated on each subindex, and guide economies from three different perspectives of competitiveness:

- GCI-1: Guides the so-called *factor-driven economy*
- GCI-2: Guides the so-called *efficiency-driven economy*
- GCI-3: Guides the so-called *innovative-driven economy.*

The pillars are explained as follows:

- *GCI-1*: *Based on Pillars 1–4*:
  - *Pillar 1*: *Institutions*—governance both in public and private sectors, possibly in the third sector—non-profit organizations (Putz, 2019a).
  - *Pillar 2*: *Infrastructure—implies the existence of the high-quality mobility facilities*, such as highways (highways, speed lanes), railways, ports, airports, all serving multinational and entrepreneurial businesses, as well as the mobility of services and work; it also requires *sustainable electricity network* running uninterruptedly and if possible, with a low environmental impact; and not less important, it needs *telecoms* as a support for information flow and real-time decision-making.

- *Pillar 3*: *Macroeconomic environment*—although it does not guarantee economic growth (or the productivity) by itself, it must be stable in order to react to potential crises, by ensuring natural cycles (durable at the firm level, while sustainable at the level of society), without being threatened by imbalances caused by governments with deficits in public balance, or by other means, etc.
- *Pillar 4*: *Health and primary education*—poor health is causing high economic costs and diminishing productivity and competitiveness; likewise, basic education ensures a level of morality in society and stimulates efficiency and motivation for progress at the individual level.

• *GCI-2*: *Based on Pillars 5–10*:

- *Pillar 5*: *Higher education and training* means continuous and advanced education that ensures economic development beyond the mere production of goods and services (processes); secondary, (tertiary) and tertiary education (higher education) should increase the capacity to understand and adapt to changes in the environmental and economic environment in a global context, and to be able to produce leaders in business, as well as on-job training, thus, ensuring a perpetual updating of employees' skills.
- *Pillar 6*: *Goods market efficiency* ensures the optimal mix of goods and services, between supply and demand, both by the country and from foreign investment. It provides goods that, in their turn, enhance prosperity (goods that thrive) and sophistication of/to customers; it also takes into account the comparative advantage between different countries, according to the cultural profile of consumers, bearing different requirements and expectations.
- *Pillar 7*: *Labor market efficiency* represents labor market flexibility, motivation and dynamics, the capacity of recognizing talent and meritocracy; it should allow workers to shift from one job, or enterprise to another, at low cost, with possible fluctuations in income that, nevertheless, do not unbalance society (taking into account the potential passages of unemployment, retraining, relocation, etc.).

- *Pillar 8*: *Financial market development* is based on population's savings, foreign entrants, entrepreneurial actions, investment projects, with an economic feedback that is superior to any related policy measures. It also refers to the banking sector with the transparent and efficient credit sector designed for the private and entrepreneurial needs, while practicing secure and advantageous transfer rates, supporting win–win business of type *joint venture,* and protecting the risk capital. The capital market must also be confined by specific measures, observing the security of investments and the guarantee of equal opportunities in an open economy.
- *Pillar 9*: *Technological readiness* measures the agility with which an economy adopts technology transfer, and offers the access to the functional/blueprint layouts, especially by means of information and communication technology. It should be active, even with purchasing and rental costs, including, but not limited to, the direct foreign investment. The upgrading of the existing daily production activities toward the most advanced level of technology, through and for leveraging, is favoring efficiency, competitiveness, innovation of product, process, the organization structure, its process flow, its market including distribution, and strategy overall.
- *Pillar 10*: *Market size* also includes the internal and external market, with an emphasis on export-based economies (productive, competitive, and specific to the European Union).

• *GCI-3*: *based on Pillars 11–12*:

- *Pillar 11*: *Business sophistication* is a postmodern feature measuring, for example, the network of firms and their degree of interconnectivity and their degree of agglomeration and co-opetition (cooperation + competition) in industrial clusters or districts, in technology parks, business incubators, and business accelerators (there where, unlike incubators, participants can invest in each other according to cooperation agreements).
- *Pillar 12*: *Innovation* measures the degree of competitiveness through technological readiness level. It is in turn based on research and development at the frontiers of scientific knowledge. Thus, an increased and/or the competitive productivity

is achieved through the added-value generated by innovation; this behavior is specific to the private sector, but with a recent tendency to involve universities and research institutes in spin-offs and/or start-ups joint ventures and projects, with the protection of intellectual property, yet enhancing their edge of competitive potential (patents, know-how, etc.).

The breakdown of these subindices and pillars on variables and subvariables, together with their cumulative hierarchical weight is illustrated in Table 10.2.

TABLE 10.2 Component of Global Competitiveness Index, GCI-2016, with Subindices, Pillars, Variables and Subvariables, Expressed as Hierarchical Cumulative Percentages; the Notification (1/2) Indicates the Retrieval of the Subvariable in Question and in Another Pillar-Subindex Variable, Respectively.

Subindex, pillar, variable, subvariable	Percent
GCI-1: Basic requirements	20–60%
Pillar 1: Institutions	25%
Public institutions	75%
Property rights	20%
1.01. Property rights	
1.02. Intellectual property protection (1/2)	
Ethics and corruption	20%
1.03. Diversion of public funds	
1.04. Public trust in politicians	
1.05. Irregular payments and bribes	
Undue influence	20%
1.06. Judicial independence	
1.07. Favoritism in decisions of government officials	
Public-sector performance	20%
1.08. Wastefulness of government spending	
1.09. Burden of government regulation	
1.10. Efficiency of legal framework in settling disputes	
1.11. Efficiency of legal framework in challenging regulations	
1.12. Transparency of government policymaking	
Security	20%
1.13. Business costs of terrorism	
1.14. Business costs of crime and violence	

TABLE 10.2 *(Continued)*

Subindex, pillar, variable, subvariable	Percent
1.15. Organized crime	
1.16. Reliability of police services	
Private institutions	25%
Corporate ethics	50%
1.17. Ethical behavior of firms	
Accountability	50%
1.18. Strength of auditing and reporting standards	
1.19. Efficacy of corporate boards	
1.20. Protection of minority shareholders' interests	
1.21. Strength of investor protection	
Pillar 2: Infrastructure	25%
Transport infrastructure	50%
2.01. Quality of overall infrastructure	
2.02. Quality of roads	
2.03. Quality of railroad infrastructure	
2.04. Quality of port infrastructure	
2.05. Quality of air transport infrastructure	
2.06. Available airline seat kilometers	
Electricity and telephony infrastructure	50%
2.07. Quality of electricity supply	
2.08. Mobile telephone subscriptions (1/2)	
2.09. Fixed telephone lines (1/2)	
Pillar 3: Macroeconomic environment	25%
3.01. Government budget balance	
3.02. Gross national savings	
3.03. Inflation	
3.04. Government debt	
3.05. Country credit rating	
Pillar 4: Health and primary education	25%
Health	50%
4.01. Business impact of malaria	
4.02. Malaria incidence	
4.03. Business impact of tuberculosis	
4.04. Tuberculosis incidence	
4.05. Business impact of HIV/AIDS	

TABLE 10.2 *(Continued)*

Subindex, pillar, variable, subvariable	Percent
4.06. HIV prevalence *(percent of population carrying the virus*	
4.07. Infant mortality	
4.08. Life expectancy	
Primary education	50%
4.09. Quality of primary education	
4.10. Primary education enrollment rate	
GCI-2: Efficiency enhancers	35–50%
Pillar 5: Higher education and training	17%
Quantity of education	33%
5.01. Secondary education enrollment rate	
5.02. Tertiary education enrollment rate	
Quality of education	33%
5.03. Quality of the educational system	
5.04. Quality of math and science education	
5.05. Quality of management schools	
5.06. Internet access in schools	
On-the-job training	33%
5.07. Local availability of specialized research and training services	
5.08. Extent of staff training	
Pillar 6: Goods market efficiency	17%
Competition	75%
Domestic competition	67%
6.01. Intensity of local competition	
6.02. Extent of market dominance	
6.03. Effectiveness of anti-monopoly policy	
6.04. Effect of taxation on incentives to invest	
6.05. Total tax rate	
6.06. Number of procedures required to start a business	
6.07. Time required to start a business	
6.08. Agricultural policy costs	
Foreign competition variable	33%
6.09. Prevalence of trade barriers	
6.10. Trade tariffs	
6.11. Prevalence of foreign ownership	
6.12. Business impact of rules on Foreign Direct Investments (FDI)	

TABLE 10.2 *(Continued)*

Subindex, pillar, variable, subvariable	Percent
6.13. Burden of customs procedures	
6.14. Imports as a percentage of GDP	
Quality of demand conditions	33%
6.15. Degree of customer orientation	
6.16. Buyer sophistication	
Pillar 7: Labor market efficiency	17%
Flexibility	50%
7.01. Cooperation in labor–employer relations	
7.02. Flexibility of wage determination	
7.03. Hiring and firing practices	
7.04. Redundancy costs	
7.05. Effect of taxation on incentives to work	
Efficient use of talent	50%
7.06. Pay and productivity	
7.07. Reliance on professional management (1/2)	
7.08. Country capacity to retain talent	
7.09. Country capacity to attract talent	
7.10. Female participation in labor force	
Pillar 8: Financial market development	17%
Efficiency	50%
8.01. Availability of financial services	
8.02. Affordability of financial services	
8.03. Financing through local equity market	
8.04. Ease of access to loans	
8.05. Venture capital availability	
Trustworthiness and confidence	50%
8.06. Soundness of banks	
8.07. Regulation of securities exchanges	
8.08. Legal rights index, *the rights to the bank disrupt*	
Pillar 9: Technological readiness	17%
Technological adoption	50%
9.01. Availability of latest technologies	
9.02. Firm-level technology absorption	
9.03. FDI and technology transfer	
ICT use (information and communication tech)	50%

TABLE 10.2 *(Continued)*

Subindex, pillar, variable, subvariable	Percent
9.04. Internet users	
9.05. Broadband Internet subscriptions	
9.06. Internet bandwidth	
9.07. Mobile broadband subscriptions	
9.08=2.08. Mobile telephone subscriptions (1/2)	
9.09=2.09. Fixed telephone lines (1/2)	
Pillar 10: Market size	17%
Domestic market size	75%
10.01. Domestic market size index	
Foreign market size	25%
10.04. Foreign market size index	
GCI-3: Innovation and sophistication factors	5–30%
Pillar 11: Business sophistication	50%
11.01. Local supplier quantity	
11.02. Local supplier quality	
11.03. State of cluster development	
11.04. Nature of competitive advantage	
11.05. Value chain breadth	
11.06. Control of international distribution	
11.07. Production process sophistication	
11.08. Extent of marketing	
11.09. Willingness to delegate authority	
11.10=7.07. Reliance on professional management (1/2)	
Pillar 12: R&D innovation	50%
12.01. Capacity for innovation	
12.02. Quality of scientific research institutions	
12.03. Company spending on R&D	
12.04. University-industry collaboration in R&D	
12.05. Government procurement of advanced technology products	
12.06. Availability of scientists and engineers	
12.07. PCT (patent cooperation treaty) patent applications	
12.08=1.02. Intellectual property protection (1/2)	

It should be remarked that the most important key to competitiveness lies in the *economic efficiency—specific to management*. It emphasizes on the *path* to achieving the objectives, with a great attention to the resource economy, plus the component of innovation in the circumstantial situations of the resource crisis, financial, personnel, political, legal, etc. In contrast to it, the *specific effectiveness of the traditionalist economy* promotes profit or achievement of objectives at any cost, placing the "care" for spending resources on the secondary level.

And it is precisely the last pillar (no. 12)—the one based on innovation, the one that, necessarily, opens the research and development (R&D) concern in the research–development–innovation triptych, as the last necessary component in considering globalization of a sustainable (inter-, multi-, and transnational) economy. Thus, some essential dimensions of emerging R&D problems are as follows:

- *Universities enhance their role as global players* (Aebischer, 2015):
  - Through global competition, in a global "family," possibly reunited through *the Academic Ranking of World Universities (ARWU)* yearbook, first published in 2003 by the *Center for World-Class Universities of Shanghai Jiao Tong University,* China, reporting the famous top 500 of 35,000 registered universities.
  - Explosive growth in brain circulation (based on a growth estimate of about 63 million more global students in 2025, up to nearly 262 million in the same year, with almost half only in India and China).
  - By the relevance of curricula, closer to the leap of innovation, being creators of new knowledge (in 2010, nearly 3 billion people lived near the innovation areas).
  - By raising awareness of intellectual property rights and engaging in reversed innovation (*bottom-up*, from fundamental to applicative).
  - Through digital disruption that finds a new way for global research (including virtual research and study networks. This includes *open-access* science, while the *massive open on-line courses* are potentiated by three factors, namely: (1) the technological advance of digitization; (2) the "digital natives," who are currently (almost as 12 million) attracted by the maintenance of these university courses, and (3) by filling the lack of access to textbook libraries.

- Enhance partnerships between universities—creation of university consortia, as regional clusters *(hard-nature),* or as global-virtual communication networks (*soft-nature*).

Development–innovation triptych, as the last necessary component in considering globalization of a sustainable (inter-, multi-, and transnational) economy. Thus, some essential dimensions of emerging R&D problems are as follows:

➢ *Inclusive science development* (Neupane, 2015):

- Through Science 2.0 program on Data Revolution (Accessibility)
- By increasing collaborative science *(big data goes global).*
- By shifting the emphasis from basic research to big science with a socioeconomic impact, by addressing the overall (bigger) picture of the human race's challenges—global pandemics, water, food and energy, insecurity, and global climate change.
- Citizen science are involved in cogenerating knowledge through the open-access culture, that is, by finding *crowd-sourcing* solutions (with the power of the enthusiastic crowd), thus forcing the so-called *data down scaling* processes (the reshaping at the individual level) and the *data mainstream* (the reshaping of interpersonal, social data flow). They, ultimately, are affecting governmental decisions through the establishment of adaptive governance. The great *highways* of scientific development are thus created: (1) the openness, (2) the inclusiveness, and (3) the respect for intellectual property—all of which lead to avoidance of duplication of research, or its faking.
- Through awareness of "information washing" at the border, data transparency and assuming the risk of uncontrolled data explosion (e.g., through the DataONE project), is potentially dangerous in the field of bionanotechnologies, astronomy, and geophysics. This way, the data governance is carried out in universities (in the traditional way) and by inter-, multi-, and transdisciplinary, in collaborative international projects (as a virtual and postmodern approach).
- Through the *commons creation community in big science,* which defines (since 2011) the governance of data as "the system of decisions, rights and responsibilities, that describe the big data

custodians and the methods used to govern them; it includes data-related regulations and policies, as well as, strategies for data quality control and their management in the context of an organization."

- By targeting the *17 sustainable development goals* (SDGs) UNESCO 2030:
- SDG-1: poverty eradication (no poverty)
- SDG-2: hunger eradication (zero hunger)
- SDG-3: good health and well-being
- SDG-4: quality education
- SDG-5: gender equality
- SDG-6: clean water and sanitation
- SDG-7: clean and affordable energy
- SDG-8: decent work and economic growth
- SDG-9: industry and innovative infrastructure (industry, innovation, and infrastructure)
- SDG-10: reducing inequalities
- SDG-11: sustainable cities and communities
- SDG-12: responsible consumption and production
- SDG-13: climate action
- SDG-14: under water life
- SDG-15: life on land
- SDG-16: peace, justice, and strong institutions
- SDG-17: partnership for these goals.

➢ *A new paradigm of global science policy* (Hackmann and Boulton, 2015):

- Through the challenge of *global change* (to be tackled by combining the criterion of sustainability with ethical actions): the occurrence of the new hazard, the impact of human action, as a geological force (anthropocen), the reconfiguration of the global ecology, of the cultural challenges (by increased migration), the decoupling of economic growth (even aiming at the so-called the *economic stasis*) from the environmental impact.
- Through the challenges of *change in science*: the so-called Future Earth agreed in 2012 by the *International Council of Science, by the International Social Science Council, UNESCO, by the United Nations Environmental Program, the World*

Meteorological Organization, by the United Nations University, and by the Belmont Forum—as a platform to promote inter- and transdisciplinary scientific practices; it aims at widening social complexity and collective solutions, including knowledge subcultures, the passive knowledge consumers, with addressability to politicians and practitioners, activists and citizens alike, in a triptych of *open data*, thus practicing the transparency of information (*open information)* and *open knowledge.*

- Through *science policy*: as a lens for the economy based on knowledge, that is, understanding the open science as a public good, or as a public action for a right and sustainable world. The science–policy interface *(science–policy nexus)* is regarded as a manifestation of *knowledge duality, that is, production and use*, causing and potentiating a multidimensional model of iterative interactions, with feedbacks and loops toward a scientific process of "messy decisions," integrated into societal stakeholders through a filter (scientific), both critical and reflexive. The task (onus) falls into this new approach of codesign, coproduction, and cocreation between academic and nonacademic environments, toward an effective policy in the global knowledge economy.

➢ *The role of local knowledge on the science–political interface* (Nakashima, 2015):

- Through the global understanding and recognition of traditional local and indigenous systems, (*indigenous knowledge*), see *The Intergovernmental Platform for Biodiversity and Ecosystem Services*—IPBES.
- Recognition of the Platonic "knowing again": the positivist (scientific) knowledge is re-gained through the *social emotion* of the need for change and adaptation, through the empiricism of the crowd combined with academic rationality and environmental objectivity (giving the resources and capacities available at a certain time). They are all conjugated to political decisions that include risks, subjectivity, hazard, and irrationality, in both local and global science–policy forum—so glocal (Robertson, 1994).

In this panorama of economic growth through direct foreign investment, competitiveness, globalization, and enhancement of open-minded science, by targeting the alignment (adaption) with the future Earth, we can move on to the global causal study on how global factors and indices correlate with the knowledge economy through direct investment in research and development—thus offering an analytical and then strategical interpretation, as it follows.

10.3 METHODOLOGICAL ANALYSIS

UNESCO Report Science: 2030 Horizon in global research–development–innovation with global–local economic impact (glocal)! Shifting from knowledge (indicators) to knowledge (development strategies) involves shaping an index of the knowledge economy in the 3D Space of the glocal economy (global science development, totally open in the economic and global sustainable development area)!

For this strategic modeling goal, M-global descriptors are considered in a methodological manner; they are considered as cumulative causes, super imposable in the resulting activity (with both global and local effect, thus glocal).

$$|Y\rangle = a_0|X_0\rangle + a_1|X_1\rangle + a_2|X_2\rangle + ... + a_M|X_M\rangle. \quad (10.4)$$

The algebraic form (eq 10.4) can be seen in the strict sense of the cause–effect when the glocal activity $|Y\rangle$ breaks down on the causal contributions given by the vectors $\{|X_0\rangle, |X_1\rangle, |X_2\rangle, ..., |X_M\rangle\}$, with the unitary vector of the free term $|X_0\rangle = \hat{1}_N = |11...1_N\rangle$, and the projection coefficients $a_0, a_1, \ldots, a_M$. However, the problem of endogenity appears, naturally, due to the fact that the vectors $\{|X_1\rangle, |X_2\rangle, ..., |X_M\rangle\}$ are not perfectly orthogonal, $|X_1\rangle \perp |X_2\rangle \perp ... \perp |X_M\rangle$, which drastically reduces the possibility of using different correlation indices, especially when they are global—due to inherent interferences between their enveloping and characterizing areas (Section 10.2). Fortunately, this limitation is removed by the original mechanism of considering the algebraic-statistical correlation (eq 10.4), even using vectors. They are variables of a generally nonorthogonal correlation, yet using a Gram–Schmidt orthogonalization mechanism this drawback can be resumed by the multilinear expansion, for instance, after the first column of the determinant equation, with the typical display (Putz and Lacrama, 2007; Putz, 2012, 2016):

$$\begin{vmatrix} |Y\rangle & \omega_0 & \omega_1 & \cdots & \omega_k & \cdots & \omega_M \\ |X_0\rangle & 1 & 0 & \cdots & 0 & \cdots & 0 \\ |X_1\rangle & r_0^1 & 1 & \cdots & 0 & \cdots & 0 \\ \vdots & \vdots & \vdots & \vdots & & \vdots & \\ |X_k\rangle & r_0^k & r_1^k & \cdots & 1 & \cdots & 0 \\ \vdots & \vdots & \vdots & \vdots & & \vdots & \\ |X_M\rangle & r_0^M & r_1^M & \cdots & r_k^M & \cdots & 1 \end{vmatrix} = 0, \tag{10.5}$$

In eq 10.5, the determinant components are calculated with the rule of the vector product for algebraic vectors of N-component

$$\omega_k = \frac{\langle \Omega_k | Y \rangle}{\langle \Omega_k | \Omega_k \rangle}, \quad k = \overline{0, M}, \tag{10.6}$$

$$r_i^k = \frac{\langle X_k | \Omega_i \rangle}{\langle \Omega_i | \Omega_i \rangle}, \tag{10.7}$$

with orthogonal intermediate vectors $|\Omega_0\rangle$, $|\Omega_1\rangle$, ..., $|\Omega_{k-1}\rangle$, ..., from the iterative construct of the Gram–Schmidt type, starting from the orthogonality condition of the first two vectors

$$\langle \Omega_0 | \Omega_1 \rangle = 0 \tag{10.8}$$

then propagated in a multilinear manner of the form:

$$|\Omega_k\rangle = |X_k\rangle - \sum_{i=0}^{k-1} r_i^k |\Omega_i\rangle, \; k = \overline{0, M}. \tag{10.9}$$

The mechanism is particularly useful for determining the causal hierarchies, which is what imposes precedence and, thus also, the influence on the global effect (scale). This can be done by considering the algorithm in eqs 10.4–10.10, initially introduced for quantitative chemical structure—biological activity studies *(namely, quantitative structure–activity relationships, QSAR),* to the present global generic indexes. This way, one can measure the influence the open and global science have in the knowledge economy, thus producing the present called *quantitative knowledge–global relationships* (QKGR)—quantitative knowledge–R&D–globalization study. In this respect, the QKGE algorithm completes with the spectral-KGR formalism, which considers the 1-, 2-, ... , M-linear correlations, for which the Euclidean paths are calculated as follows:

$$\delta[l,l'] = \delta\sqrt{\left(\left\||Y_l\rangle\right\| - \left\||Y_{l'}\rangle\right\|\right)^2 + \left(R_l - R_{l'}\right)^2},\ l,l'\!:\!ENDPOINTS\ \ MODELS \quad (10.10)$$

$$\left\||Y\rangle^{ENDPOINT}\right\| = \sqrt{\sum_{i=1}^{N} y_i^2} \quad (10.11)$$

between all combinations of different correlation levels (mono-, bi-, multi-variables) and the *minimum path* in the series (spectrum) of these data is selected

$$0 = \delta\left\{|Y_{Ii}\rangle, |Y_{IIi}\rangle, \ldots, |Y_M\rangle\right\}_{i=\overline{1,M}}. \quad (10.12)$$

The spectral development of the eq 10.13 will indicate the causal hierarchy of the global variables indices that come into the glocal economy of knowledge (R&D), with the possibility of identifying competitive, sustainable, and regenerative development strategies.

The following application uses the numerical global indices presented in this study, at the level of the European Union (Table 10.3).

The results of applying the QKGB algorithm in the spectral-KGR variant, with eqs 10.4–10.13 for generating, classifying, and interpreting knowledge correlations through R&D with economic globalization, are presented in Tables 10.4–10.6, and facilitate some interesting observations, both statistically and predictively, as follows:

- From Table 10.4, we can observe from the total correlation, model III with all the included indicator variables, how the indicators $|X_2\rangle = KOF$ and $|X_3\rangle = WIR$ contribute very little to the activity $|Y\rangle = R\&D$, as compared to the priority contribution of competitiveness through the associated index $|X_1\rangle = GCI$.
- From Table 10.5, it is confirmed that the R&D–competitiveness correlation has the highest (Pearson) correlation coefficient of the one-core models (the model Ia). However, also, in the bi-correlation combinations one remarks IIa<IIb, along with the KOF indices IIa and investment in model IIb, thus being the last model found with the highest correlation factor—toward the aggregated correlation in the model III.
- Moreover, Table 10.5 results in a very interesting fact—namely, that the *algebraic vector ("size") of the R&D in the studied series, for*

TABLE 10.3 Numerical Values of Employed Global Indices: the UNESCO-R&D as Prospected Toward 2030 (the Target is in Percentages of the National GDP), According to Hollanders and Kanerva (2015); the Same with Indices WIR-2013 (WIR, 2015), GCI 2015 (GCI, 2015), and KOF-2013 (KOF, 2016). They are Processed in the Spectral-KGR Algorithm for Seeking the Correlation of Open Knowledge with Globalization in the Context of the Knowledge Economy; See the Text for Details at the Level of the European Union (28 Countries).

Country	R&D (% GDP)	Free term	GCI-2015	KOF-2013	WIR-2015
Algebraic vectors	$\|Y\rangle$	$\|X_0\rangle$	$\|X_1\rangle$	$\|X_2\rangle$	$\|X_3\rangle$
Austria	3.76	1	5.12	89.82929	125
Belgium	3.00	1	5.20	90.50983	66
Bulgaria	1.50	1	4.32	77.15737	132
Croatia	1.40	1	4.07	75.59323	121
Cyprus	0.50	1	4.23	84.07279	92
Czech Republic	3.00[a]	1	4.69	83.5984	143
Denmark	3.00	1	5.33	86.44198	119
Estonia	3.00	1	4.74	78.46434	92
Finland	4.00	1	5.45	85.467	136
France	3.00	1	5.13	82.61184	167
Germany	3.00	1	5.53	78.24411	198
Greece	0.67	1	4.02	80.40121	107
Hungary	1.80	1	4.25	85.7771	122
Ireland	2.00	1	5.11	91.64422	64
Italy	1.53	1	4.46	79.58942	155
Leetonia	1.50	1	4.45	70.96922	108
Lithuania	1.90	1	4.55	77.25558	118
Luxemburg	2.60	1	5.20	83.55183	157
Malta	0.67	1	4.39	75.03847	86
Holland	2.50	1	5.50	91.69517	160
Poland	1.70	1	4.49	79.89501	126
Portugal	3.00	1	4.52	85.07809	119
Romania	2.00	1	4.32	75.08933	146
Slovakia	1.20	1	4.22	83.62099	119
Slovenia	3.00	1	4.28	76.23893	101
Spain	2.00	1	4.59	83.72595	146
Sweden	4.00	1	5.43	85.92312	133
UK	3.00[a]	1	5.43	81.96537	168

[a]In Czech Republic and the UK—note that due to the Brexit Decision of 2016, there is no R&D target statement for UK, yet it is fixed at the EU-2020/2030 average.

the countries of the European Union, go at the rate over 12% GDP in all models. This indicates the actual influence (of the current and the forecast, even slightly higher, with the maximum in the total model III) of R&D knowledge in the European Knowledge Economy (and even higher, to the average of the national GDP assumed at the level of the European Union of about 3%); fortunately, this is another expression of the synergistic effect of R&D in the "cluster" of European nations—the European Union. The meaning and the significance of the phrase "knowledge economy" is, therefore, fully assumed as a desideratum at European Community level.

TABLE 10.4 The Knowledge–Globalization Equations Modeled with the Spectral-KGR Algorithm Based on Eqs 10.4–10.9 for All Correlation Combinations Using the Data in Table 10.3.

Models	Vectors	Equation "Spectral-KGR"
Ia	$\vert X_0\rangle, \vert X_1\rangle$	$\vert Y\rangle^{Ia} = -4.82004\,\vert X_0\rangle + 1.49745\,\vert X_1\rangle$
Ib	$\vert X_0\rangle, \vert X_2\rangle$	$\vert Y\rangle^{Ib} = -3.85764\,\vert X_0\rangle + 0.0749066\,\vert X_2\rangle$
Ic	$\vert X_0\rangle, \vert X_3\rangle$	$\vert Y\rangle^{Ic} = 1.12418\,\vert X_0\rangle + 0.00928895\,\vert X_3\rangle$
IIa	$\vert X_0\rangle, \vert X_1\rangle, \vert X_2\rangle$	$\vert Y\rangle^{IIa} = -4.77191\vert X_0\rangle + 1.5028\,\vert X_1\rangle - 0.000895642\,\vert X_2\rangle$
IIb	$\vert X_0\rangle, \vert X_1\rangle, \vert X_3\rangle$	$\vert Y\rangle^{IIb} = -4.83265\,\vert X_0\rangle + 1.48202\,\vert X_1\rangle + 0.00068243\,\vert X_3\rangle$
IIc	$\vert X_0\rangle, \vert X_2\rangle, \vert X_3\rangle$	$\vert Y\rangle^{IIc} = -5.63648\,\vert X_0\rangle + 0.0803801\,\vert X_2\rangle + 0.0105563\,\vert X_3\rangle$
III	$\vert X_0\rangle, \vert X_1\rangle, \vert X_2\rangle, \vert X_3\rangle$	$\vert Y\rangle^{III} = -4.87877\,\vert X_0\rangle + 1.47589\,\vert X_1\rangle + 0.000841\,\vert X_2\rangle + 0.000731303\,\vert X_3\rangle$

TABLE 10.5 Spectral-KGR Spectral Rule Reporting by Eqs 10.10 and 10.11 and Statistical Binding Correlation for All Correlation Patterns in Table 10.4, for the Input Data in Table 10.3, Respectively.

	Ia	Ib	Ic	IIa	IIb	IIc	III
$\left\Vert \vert Y\rangle^{PREDICTED} \right\Vert$	12.735	12.3182	12.2296	12.735	12.7354	12.4333	12.7354
$R^{STATISTIC}_{Spectral-KGR}$	0.75645	0.411808	0.29285	0.756461	0.756716	0.52862	0.756724

- Table 10.6 shows the hierarchy of intercorrelation paths. It "says" about the "most probable/optimal, thus strategic" successive movements to follow, in order to achieve the maximum correlation. In the current context of about 75% of knowledge, R&D based on the globalization is well represented by the GCI, KOFG, and WIR indices; it is worth mentioning that, "at equality" (or degeneration, in statistical terms), for equal distances on different paths, the minimum path will be established by considering the highest statistical correlation in a couple of successive models with the same starting "point"; accordingly, for the "Ia" starting model, the model IIb has a correlation factor, superior to model IIa, where from, although aligned on global paths Ia-IIa-III and Ia-IIb-III, the latter path is considered as the preferred inter-model correlation "wave," originating in the Ia model. It is then continued with the search for the minimum path for the rest of configurations originating in the Ib and Ic models, establishing (without degeneration this time) "the second and third wave," the so-called β and γ paths.

TABLE 10.6 Inter-Models Spectral-KGR Paths in Table 10.4, Calculated with the Data in Table 10.5 and Eq 10.11, with the Shortest Distance and Hierarchy for the First Shortest Paths According to the Minimum Route Procedure (10.12), See also the Text for Details.

Path	Distance	Hierarchy
Ia-IIa -III	0.000503757	
Ia-IIb-III	**0.000503757**	α
Ia-IIc-III	0.756658	
Ib-IIa-III	0.541366	
Ib-IIb-III	0.542583	
Ib-IIc-III	*0.541363*	β
Ic-IIa-III	0.686334	
Ic-IIb-III	0.686327	γ
Ic-IIc-III	0.69015	

The global influence in the QKGR correlation present (on the data in Table 10.3) suggests the (strategic) path of the correlations

$$Ia \rightarrow IIb \rightarrow III \tag{10.13}$$

which, rewritten in terms of the variables involved in the correlation paths, according to Table 10.4, results in:

$$|X_1\rangle \rightarrow |X_1, X_3\rangle \rightarrow |X_1, X_2, X_3\rangle, \tag{10.14}$$

hierarchy equivalent to the selection of essential correlation variables (by successive deletion of the selected variable in the previous hierarchical step/model)

$$|X_1\rangle \to |X_3\rangle \to |X_2\rangle, \tag{10.15}$$

and which is further transposed (see the correspondence of the vector inputs in Table 10.3 into the hierarchy of the causes for determining glocal activity of R&D, that is:

$$|GCI\rangle \to |WIR\rangle \to |KOF\rangle. \tag{10.16}$$

The final result is that in the glocalization (globalization, clustering, respecting local traditions, and values) of R&D, global factors fall into causative determinants, in the order of:

$$\text{R\&D}: \text{Competitivity} \to \text{Investments} \to \text{Productivity}. \tag{10.17}$$

Current spectral-KGR analysis was performed at the global index level, so that it is sustainable in the social environment. Instead, it also allows a phenomenological (re)interpretation at the firm level, in the light of the *competitive, sustainable, and regenerative advantage* by rescaling the paths into the distinctive advantage cube (Putz, 2017, 2019a, 2019b), as indicated in the conclusions and perspectives section.

10.4 CONCLUSIONS AND PERSPECTIVES

Thus, on the *dual nature* of globalization indices, it is concluded (Martens et al., 2015):

- They measure objectively (quantitatively)
- They measure constructively, by their composite nature.

On the addressability of globalization indices, it is concluded (Scholte, 2005; Beck, 2004; Sassen, 2000):

- The concept of globalization (what it includes, and thus defines): internationalization, liberalization, universalization versus Occidentalization, de territorialization, overterritorialization, etc.
- What field and economic, ecological, cultural, political, or R&D phenomena they capture

- The units used and the statistic correlation mechanism (De Lombaerde and Lapadre, 2008) in the country paradigm (geography and economy) + nation (community and culture) + state (governance and politics) = society.

On the multiscale approach of the globalization issue, it is concluded: *individual* → *city* → *country* → *region* → *planet* → ..., on both micro- and macroeconomic (exchange goods) and strategic (paths/exchange processes) levels.

However, the so-called *Lucas paradox* (1990) appears: why, if in the neo-classical theory of production functions (which promote the diminution of the marginal productivity of capital)—this capital should flow from the capital-rich countries to the deficient ones, but the flow of capital is predominantly oriented toward rich countries?

A possible solution can be achieved by exploiting the results of the spectral-KGR methodology for the orientation of the strategic paths (where from the "natural flow" of the capital) in the cube of the distinctive advantage in business (see also Chapter 4 of the present monograph), with the extension to the competitive advantage of the nations.

Thus, in the distinctive advantage cube (Putz 2017, 2019a, 2019b), one should have as a reference for further comparison the "reverse" path from the "red ocean" business of "animal spirits" to the "wise business." It is an algebraic chain of the longest path (Putz, 2018b) combining the coordinates (1) of competitivity (on 0X here modeled by GCI), (2) of sustainability (on 0y, here, modeled by the KOF factor), and (3) of regeneration (on 0Z, here, modeled by the WIR factor) with the minimum entropy (Fig. 10.1), namely:

$$\begin{aligned}&(-1\ \ -1\ \ -1)\big|_{\substack{ANIMAL\\SPIRITS}} \to (1\ \ -1\ \ -1)\big|_{POLEMOCRACY} \to (1\ \ 1\ \ -1)\big|_{\substack{SMART\\BUSINESS}}\\&\to(-1\ \ 1\ \ -1)\big|_{OUTSOURCING} \to (-1\ \ 1\ \ 1)\big|_{\substack{UNDERGROUND\\BUSINESS\\(KLEPTOCRACY)}} \to (-1\ \ -1\ \ 1)\big|_{\substack{PERVERSE\\SPIRITS}}\\&\to(1\ \ -1\ \ 1)\big|_{\substack{PROFITABLE\\BUSINESS}} \to (1\ \ 1\ \ 1)\big|_{\substack{WISE\\BUSINESS}}\end{aligned} \tag{10.18}$$

On the other hand, one just founded that in the presence of the R&D effect, as a result of all these causes, one deals with the correlation on the shortest path, as specific to the business economy in the knowledge (open) era. For it, the direct correspondence is achieved, according to the path (eq 10.16) projected on the coordinates 0X, 0Z, and 0Z, respectively, so we have (Fig. 10.2):

$$\begin{aligned}&(-1\quad -1\quad -1)\big|_{\substack{ANIMAL\\SPIRITS}} \rightarrow (1\quad -1\quad -1)\big|_{POLEMOCRACY}\\&\rightarrow (1\quad -1\quad 1)\big|_{\substack{PROFITABLE\\BUSINESS}} \rightarrow (1\quad 1\quad 1)\big|_{\substack{WISE\\BUSINESS}}\end{aligned} \tag{10.19}$$

This approach is coherent and consistent, as it manages to select from the (longest) strategic road the shortest one, that maximally correlates (with the QKGR model) global causes with the glocal effect of open-R&D in knowledge economy.

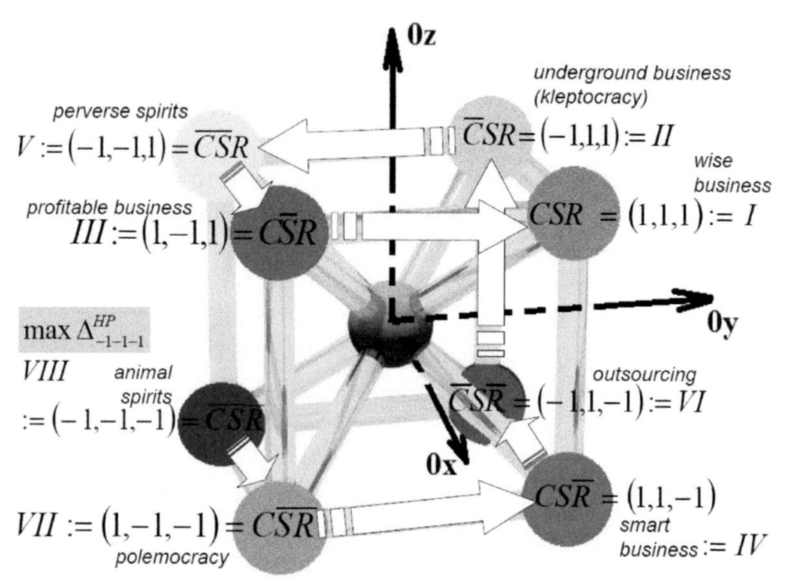

FIGURE 10.1 The maximal path (with minimal entropy) from "animal spirits business" to "wise business" in the strategic cube of distinctive advantage.
Source: Adapted with permission from Putz (2017, 2019b).

However, one remains to investigate, in a later study, on the cause–effect behavior in solving endogenous problem; and this is in relation to the Lucas paradox of capital flow in the glocal phenomenology of open-label research and development. It can make use, for example, on the contextual clusters in economic and innovation parks—yet, being in conjunction with the global (network) connectivity of global business. The clustering paradigm has the advantage of globally managing the chain of production, from suppliers to customers, bearing the mark of competition and moving within the ecolo–politico–social environment)! Thus, the glocal result can be materialized, both strategically and quantitatively, in the predicatively postmodernism (transparent, inclusive, and facilitating) of knowledge economy!

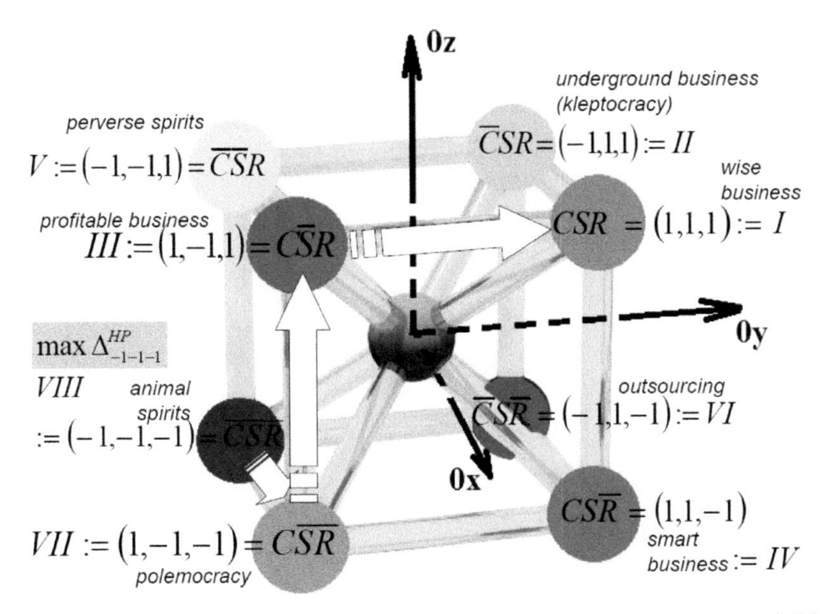

FIGURE 10.2 The minimal path (with the highest correlation in the current Spectral-KGR approach) from "animal spirits type business" to "wise business" in the strategic cube of distinctive advantage.

KEYWORDS

- **eco-political-social index**
- **world investments index**
- **global competitiveness index**
- **modeling through knowledge-globalization relationships**
- **spectral knowledge-globalization relationship**
- **strategic cube of the distinctive advantage**

REFERENCES

Aebischer P. Universities: Increasingly Global Players. In *UNESCO Science Report. Towards 2030*, 2015, pp 3–5.

Agénor, P. R. Does Globalization Hurt the Poor? *Int. Econ. Econ. Policy* **2004,** *1* (1), 21–51.

Almond, G.; Verba, S. *The Civic Culture*; Princeton University Press: Princeton, 1963.

Aron, R. *Introduction to the Philosophy of History: An Essay on the Limits of Historical Objectivity*; Beacon Presss: Boston, 1961.

Beck, U. *Der kosmopolitische Blick Order: Krieg ist Frieden*; Suhrkamp Verlag: Frankfurt am Main, 2004.

Berggren, N.; Nilsson, T. Globalization and Transmission of Social Values. The Case of Tolerance. *J. Compar. Econ.* **2015,** *43*, 371–389.

Bergh, A.; Nilsson, T. Is Globalization Reducing Absolute Poverty? *World Dev.* **2014,** *62*, 42–61.

Boix, C. *Democracy and Redistribution*; Cambridge University Press: New York, 2003.

Caselli, M. *Trying to Measure Globalization. Experiences, Critical Issues and Perspectives*; Springer: Dordrecht, 2012.

Chotikapanich, D.; Griffiths, W. E.; Prasada Rao, D. S.; Valencia, V. Global Income Distributions and Inequality, 1993 and 2000: Incorporating Country-Level Inequality Modeled with Beta Distributions. *Rev. Econ. Stat.* **2012,** *94* (1), 52–73.

DHL Global Connectedness Index, 2014. https://www.dpdhl.com/en/media-relations/specials/global-connectedness-index.html

Dreher, A. Does Globalization Affect Growth? Evidence from a New Index of Globalization. *Appl. Econ.* **2006a,** *38*, 1091–1110.

Dreher, A. The Influence of Globalization on Taxes and Social Policy: An Empirical Analysis for OECD Countries. *Eur. J. Political Econ.* **2006b,** *22* (1), 179–201.

Dreher, A.; Gaston, N. Has Globalization Increased Inequality? *Rev. Int. Econ.* **2008,** *16* (3), 516–536.

Figge, L.; Martens, P. Globalization Continues: The Maastricht Globalization Index Revisited and Updated. *Globalizations* **2014,** *11* (6), 875–893

Gassebner, M.; Lamla, M. J.; Vreeland, J. R. Extreme Bounds Democracy. *J. Conflict Resolut.* **2012,** *57* (2), 171–197.

GCI—Global Competitiveness Index. *The Global Competitiveness Report* by World Economic Forum, 2016. http://reports.weforum.org/global-competitiveness-report-2015-2016/.

Gleditsch, K. S. *All International Politics Is Local: The Diffusion of Conflict, Integration, and Democratization*; University of Michigan Press: Ann Arbor, 2002.

Hackmann, H.; Boulton, G. Science for a Sustainable and Just World: A New Framework for Global Science Policy? In *UNESCO Science Report. Towards 2030*; 2015, pp 12–14.

Heinemann, F. Does Globalization Restrict Budgetary Autonomy? A Multidisciplinary Approach. *Intereconomics* **2000,** *35*, 288–198.

Hollanders, H.; Kanerva, M. European Union. In *UNESCO Science Report. Towards 2030*; 2015, pp 231–271.

Huntington, S. P. *Political Order in Changing Societies*; Yale University Press: New Haven, 1968.

Kearney, A. T. Measuring Globalization. *Foreign Policy* **2001,** *122*, 56–65.

Kluver, R.; Fu, W. Measuring cultural globalization in Southeast Asia. In *Globalization and Its Counter-forces in Southeast Asia;* Chong, T., Ed.; ISEAS Publishing: Singapore, 2008, pp 335–358.
KOF Index of Globalization, 2016. https://kof.ethz.ch/en/forecasts-and-indicators/indicators/kof-globalisation-index.html
Lipset, S. M. Some Social Requisites of Democracy: Economic Development and Political Legitimacy. *Am. Political Sci. Rev.* **1959,** *53*, 69–105.
Lockwood, B.; Redoano, M. The CSGR Globalisation Index: An Introductory Guide. *CSGR Working Paper* 155/04, 2005.
Lucas, R. E. Why Doesn't Capital Flow from Rich to Poor Countries? *Am. Econ. Rev.* **1990,** *80* (2), 92–96.
Martens, P.; Caselli, M.; De Lombaerde, P.; Figge, L.; Scholte, J. A. New Directions in Globalization Indices. *Globalizations* **2015,** *12* (2), 217–228.
Nakashima, D. Local and Indigenous Knowledge at the Science–Policy Interface. In *UNESCO Science Report. Towards 2030*, 2015, pp 15–17.
Neupane, B. A More Developmental Approach to Science. In *UNESCO Science Report. Towards 2030*, 2015, pp 6–11.
Petrişor, I. I. *The Strategic Management. The Potentiological Approach* [Originally in Romanian as: *Management Strategic. Abordare potențiologică*]; Brumar Publishing House: Timişoara, Romania, 2007.
Potrafke, N. The Evidence on Globalization. *World Econ.* **2015,** *38* (3), 509–552.
Przeworki, A. Democracy as an Equilibrium. *Publ. Choice* **2005,** *123*, 253–273.
Przeworki, A.; Alvarez, M.; Cheibub, J. A.; Limongi, F. *Democracy and Development: Political Regimes and Economic Well-being in the World*, 1950–1990; Cambridge University Press: New York, 2000.
Putz, M. V.; Lacrămă, A. M. Introducing Spectral Structure Activity Relationship (S-SAR) Analysis. Application to Ecotoxicology. *Int. J. Mol. Sci.* **2007,** *8*, 363–391.
Putz, M. V., Ed. *QSAR & SPECTRAL-SAR in Computational Ecotoxicology*; Apple Academics, CRC Press: Toronto, NJ, 2012.
Putz, M. V. *Quantum Nanochemistry: Vol V. Quantum Structure-Activity Relationship (Qu-SAR)*; Apple Academic, CRC Press: Toronto, Canada, NJ, USA, 2016.
Putz, M. V. *Strategic Cube of the Organization Competitive Advantage.* [Originally in Romanian: *Cubul strategic al avantajului competitiv al organizațiilor*]. MBA Thesis, Faculty of Economy Science and Business Administration; West University of Timişoara, 2017.
Putz, M. V. *Strategic Innovation in the Organization Governance. The 8-folding of the Mission Balance*, Chapter 9 of the Present Monograph, 2019a.
Putz, M. V. *Strategic Innovating Paths for the Distinctive Advantage. The Changing Management faraway from Equilibrium*, Chapter 4 of the Present Monograph, 2019b.
Robertson, R. Globalization or Glocalization? *J. Int. Commun.* **1994,** *1* (1), 33–52.
Sassen, S. New Frontiers Facing Urban Sociology at the Millennium. *Br. J. Sociol.* **2000,** *51* (1), 143–159.
Scholte, J. A. *Globalization. A Critical Introduction,* 2nd ed.; Palgrave: Basingstoke, 2005.
Stiglitz, J. *Globalization and Its Discontents*; W. W. Norton: New York, 2002.

UNIDO Connectedness Index, 2016. https://www.unido.org/our-focus/cross-cutting-services/partnerships-prosperity/networks-centres-forums-and-platforms/networks-prosperity/2011-connectedness-index

Vujakovic, P. How to Measure Globalization? A New Globalization Index (NGI). *FIW Working Paper* 46, 2010.

WIR—World Investment Report. *United Nations Conference on Trade and Development: UNCTAD*, 2015. http://unctad.org/en/pages/DIAE/World%20Investment%20Report/WIR-Series.aspx.

CHAPTER 11

Cubic Management of Inclusive Scientific Change

ABSTRACT

The main (what–how–when?) axes of systematic organizational changing dimensions are joined to three sets of inclusive cubic developments, to reveal eight stages of changing strategies. They assure the compact strategic and, thus, structural development, integrally covering the 3D space on the axes taken into consideration, by generating and multiplying the strategic cube unity by eight vertices. Within the cubic, strategic changing of each stage is identified by its intrinsic significance as being associated to "strong" direction (parallel) to one of the axes of change (i.e., on 0X, 0Y, and 0Z), respectively, to a dual/mixed/fuzzy combination of these ones (i.e., XY0, 0YZ, and X0Z), and to the beginning and final synergetic points of a cycle of changing (i.e., 000 and XYZ). These points/condition/stages of a strategic change are further exemplified, to the need of technological update and/or alignment, by the scientific management, potentially applied to a National Institute of Research and Development (NIRD). More specifically, one may work for the NIRD in Timişoara, Romania, as a paradigmatic case, showing for each stage the specific content, with chosen activities meant to ensure the successful development of that stage; these activities are considered as essential elements to be followed by the executive manager, or by the changing agent (proffered the scientific director) that may closely follow the management of this complex process of an R(esearch), D(evelopment), and I(nnovation) organizational development. With this occasion, the generalized 3D model of the Kotter's dual (2D network) model of an organization is also advanced.

Motto:

"When the [*mode of the*] *music changes, the walls of the city shake*!"

—Plato (2016).

11.1 INTRODUCTION

In postmodern management, the concept of inclusivity comes "together" with the one of competition. A step further, the concept of competitivity is generated as a "soft" form of the product–market relations (Kotler, 1991); it aims at adding the necessity of optimization and continuous development of this relation, in a sustainable way. Such goal may be achieved by preserving the resources (natural, raw materials, and subproduct), while integrating them into new cycles of life of the same product or in related subproducts, facilitating horizontal communication in the local, global, and global economic network, and consequently, in clusters, and meta-clusters. Shortly:

- inclusive means internal–external–continuous
- inclusive means aligning to the logic of included third (*tertium datur*)
- inclusive means tripartite, three-dimensional, and three times fundamental.

Methodologically, the tripartite approach means evolving in a three-dimensional (also conceptual) space. This way, the linear approaches (within the fashioned *value chain*) are generalized to the square/matrix approaches (the so called value plans or *value maps*); when these two, 1D and 3D, value measures are further combined to generate the cubic space (so creating the *value constellation*). In the 3D space of competitiveness, the co-opetition (co-operation and competition) prevails: it is based on sharing value (namely, the *shared value*) and on all the inclusive forms of production, dissemination, and feedbacks. The resulting economic acts are made in the open, integrator, recycling, regenerable, therefore ecological, and consequently sustainable (so having also a future)—space!

This entire complex mechanism would not be possible without managing the change, deliberately, to avoid misinterpreting lack of organization in front of economic, ecological, and social crises, inevitable to any cycle of life (organizational, of market, generational, or of national and transnational governing). Therefore, the management of change appears as a necessity to ensure a sustainable economy, an inclusive economical life, and responsible organizational governance.

11.2 COGNITIVE ANALYSIS

Generally, change is both cause and effect of evolution, but especially in science and economy:

- it is the cause of evolution in its form of *strategic potential*
- it is the effect of evolution in its form of *dynamic strategy*.

From the postmodernist point of view, change can be approached in a tripartite way, as it follows:

1. as a *continuous transition* (*step-by-step*), from one stage to another, in controlled transformations (self-controlled), or undeliberately (imposed by the external–internal environment of an organization)
2. as a *development opportunity*, an opportunity of progress, training in organizational efficiency, productivity, and profitability
3. as *an integrated strategy* to come closer to its mission, per cycles of productive and profitable competitivity, both for internal organizational environment and for the external social, cultural, and economic ecological environment.

One arrives at the system of the three dimensions, in the logic of the included third (namely, the *tertium datur*) of considering the organizational change. It is worth observing that the modeling of change management, on superior dimensions than 3D becomes tautological, redundant, and self-correlated, since the orthogonal existence of a *3D conceptual space of the management of organizational change,* already, covers the organizational change through the answers to the fundamental questions (Predişcan, 2004) as follows:

- On axis 0X: **what** to change organizationally?
- On axis 0Y: **how** to change organizationally?
- On axis 0Z: **when** to change organizationally?

Planning the abovementioned three approaches of the management of change on the three fundamental axes of understanding generates the *strategic cube of the management of change,* consisting in eight identified stages:

- *stages of strong changes*—identifies the points/stages of organizational change on its fundamental axes (the "most powerful" answer to the questions what/how/when?), in a number of three, namely, one on each axis 0X, 0Y, and 0Z
- *stages of mixed changes (fuzzy)*—identifies three points/stages of change on the plans defined as two-by-two, fundamental axes of organizational change, by the most proper answer, suitable, for two of the three fundamental questions of change, namely, on the plans XY0, X0Z, and 0YZ
- *stages of synergetic changes*—identifies the starting and the final point (in binary couples) of a cycle of change, while answering fully simultaneously to the three fundamental answers. Nevertheless, by their correlation (union), either in the point of business (or organizational) space origin 000 as in the point XYZ on the space generator, the last being along the diagonal of the primitive/unitary cube of the change management, generates (or may generate) another cycle of organizational change.

The strategic cube of the change management is shown in Table 11.1, with all of its stage, essentially described for an organizational life cycle change. The changing stages are approached, either step-by-step, type (Newton, 2009), or as an opportunity of development/progress (Bruksos, 2010); or, as an integrated strategy, internal and external, respecting the organizational environment (Putz, 2017, 2019), while following the 3D space with fundamental axes of change (0X = what?, 0Y = how?, 0Z = when?), see the representation in Figure 11.1.

Observe that, here, the fundamental temporal axis was chosen to be "upward"/0Z direction, in harmony with "time flow" that "bears" the planar mundanity in successive "layers" of its evolution, as displayed on the film framework. Each frame may seem static, when separately taken; yet, in dynamics they are taken as subsequent in a group of actions. It is here that, a difference, opposite to the suggested order of questions in an economic approach appears (Predișcan, 2004), where the [how?] action counts, "in the end," in a changing process, so that to ensure reaching the

TABLE 11.1 The Eight Stages/Phases of Organizational Change, Comparatively, in the Cube of Strategic Management.

Stage	Axis of organizational change (Predișcan, 2004)	Change STEP BY STEP (Newton, 2009)	Change OPPORTUNITY (Bruksos, 2010)	Change by the CUBIC INTEGRATED MANAGEMENT (Putz, 2017)
I	WHAT–HOW–WHEN? (000)	UNDERSTANDING OBJECTIVE: *OF CHANGE Radical or moderate change?*	HYPOTHERMIA: *do we survive or live?*	OPPORTUNITY *OF CHANGE*: *understanding internal–external context (SWOT diagnosis)*
II	WHAT–WHEN? (X0Z)	BUILDING UP THE TEAM FOR CHANGE: *identification of resources and of common essential roles* (*guarantee of change*)	DISCOMFORT: *Re-writing the scenario!*	POTENTIAL: *OF CHANGE closing preparing the opening*
III	WHAT–HOW? (XY0)	PLANNING: THE CHANGE *apologia for action, implementation of immediate gains*	ANALYSIS: *confrontation with own disorder*	SURPLUS VALUE OF CHANGE: *developing added value*
IV	HOW? (0Y)	CAPACITY FOR A CHANGE: *degree of change acceptance* (*predictability, management of reactions*)	EXPLORATION: *the well-established targets aim at the results and not to the activities; avoiding management by exception*	PROMOTING COMPETITION: *understanding and motivation of competition as a synergy of competition and cooperation*
V	WHEN? (0Z)	CHANGE IMPLEMENTATION: *testing solutions, management of risk*	EXPERIMENTATIO: *self-projection, avoiding delay, emergency versus importance of decisions*	GETTING OUT OF SELF : *controlled/oriented communication within the network*
VI	HOW–WHEN? (0YZ)	CONSOLIDATIO OF CHANGE: *steadiness of performances, recognizing errors*	DECISION: *moment of truth, acceptance of the necessity of change, checking list, keeping a good organization and "promise"!*	AMPLITUDE OF TRANSITION: *anatomy of evolution, value and increase chrysalides value (as an intermediate stage, type nymph /doll between caterpillar and butterfly)*

TABLE 11.1 *(Continued)*

Stage	**Axis of organizational change (Predişcan, 2004)**	**Change STEP BY STEP (Newton, 2009)**	**Change OPPORTUNITY (Bruksos, 2010)**	**Change by the CUBIC INTEGRATED MANAGEMENT** (Putz, 2017)
VII	WHAT? (0X)	MANAGEMENT OF COMMUNICATION: *making key message (informal and un structured in- and out-information), but also continuous training (events and formal communication), with minimization of risks of change*	COMMITMENT: *Motivation of the continuity of change, over passing resistance, acceptance of responsibility acceptance of change*	SHARING VALUES: *Value chain, closing intra-organizational circle*
VIII	WHAT–HOW–WHEN? (XYZ)	PREPARING FUTURE: *continuous change, durability of change, organizational restructuring*	RETURNING: *the change—a continuous process*	EXPORTER OF SUSTAINABILIT: *value constellation, opening that prepares reclosing*

Source: Adapted with permission from Putz (2017, 2019).

objectives (namely, substantive paradigm of economy); however, in the world in process the "time remains in the end, it is the first and the last reality to remain, or lose"; it is also a regenerator of life cycles, or the invisible, yet universal agent, binding the product with its market (Kotler, 1991). So, time is defining and determining the continuity in the changing management, viewed as the phenomenological process paradigm.

Figure 11.1 shows the complex changing management as a process in 3D space of the business or organizational strategy. The apparent "hoping," from one strategic change stage to another, may appear as jumps, only, for an external perception of the organizational change. Where from, the (mis) concept of technological and scientific *revolution* in economy, a phrase used only by the ones placed somehow distant, or exterior, to the technological and scientific activity. Instead, in fact, changing management stages are part of a complex continuity (e.g., look to the continuous path in Figure 11.1, as to a Moebius curve); thus, it describes, more correctly, the *evolutionary* process of inclusive change (when the process is deliberate, controlled, and oriented), as a pro-active process (different from the revolutionary one, which is characterized by jumps, specific for crises, and to reactive strategy only).

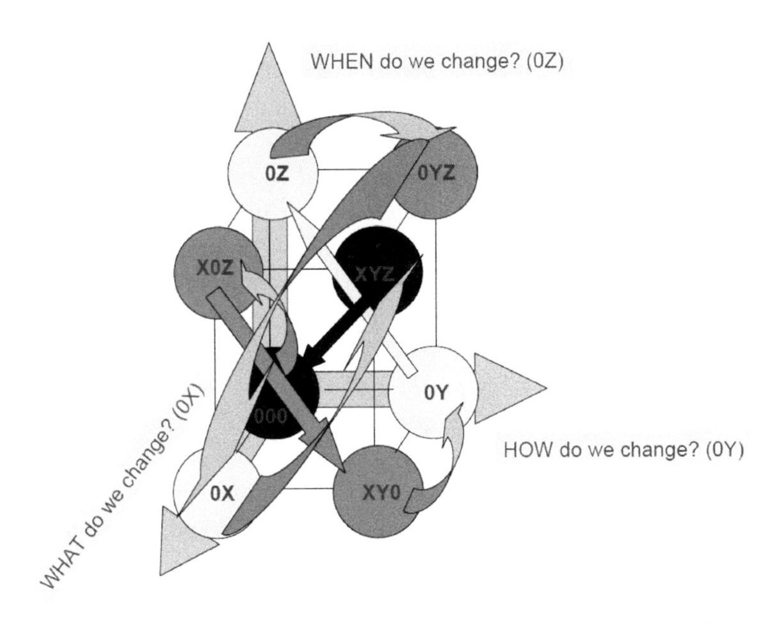

FIGURE 11.1 Strategic cube of the management of change (developed in the order of the stages in Table 11.1).

In the next section, the present approach refers to the application of inclusive management of change to the case of NIRD, Timişoara, from the prospect of its scientific and technical strategic objectives.

11.3 METHODOLOGICAL ANALYSIS

The contextual metaphor of changing: On a stirred ocean, in the North Atlantic, a multiutility ship sails, almost drifting, through the monstrous glaciers, which threatens the entire cargo (merchandises, crew, and passengers) on a terrible sinking, in a similar way with the Titanic destiny, more than 100 years ago. Captain Fred, on its senior's maturity, with gallons, with white hair under the handsome hat, with impeccable race beard empire stile, is calling impatiently his officers on report. He needs to make a saving decision, both on short term (saving the precious cargo and human) and on long term (saving the ship itself and the trust in company, etc.). The officers hastily report: the ship owners will want the ship intact, the manufacturers will want the goods to reach their destination, the passengers will want to remain safe and with minimum of emotional experiences (possibly fatal for some), the engineers from the engine chamber are already exhausted, the ocean will be polluted by the merchandises spilled all-over, the insurers will go to trial with the parent company for natural cause versus navigation error, the crew itself will be held responsible and fired in case of shipwreck with any type of damage, the area will be invaded by multinational agencies in order to dispute the investigation primacy, the company will no longer manufacture this type of ships, the touring route will be suffering, factories and companies will be closing down on the entire distribution chain for ship maintenance as well as on touristic and commercial ones, etc. What needs to be done first? What and who has priority? What is the navigation plan in case of crisis? How do we get back on the initial course and how do we get into calm waters? Who will be assuming the major decisions, solidary with the Captain? Are we following the imminent shipwreck protocol or are we dynamically adapting on the situation by risking subsequent penalties? How do we maximize the results with minimum of costs and loses?

The current strategy is an option for a long- and medium-term development (for the organizational health capable of transforming itself). Nevertheless, we need to stay with an "*occhio aperto*"/*an open eye* on the short term of the socioeconomic environment, fulfilling, at least, partially the

PESTEcJ(C) interests (political, economical, social, technical, ecological, and juridical, possible also cultural). In order to undertake a sustainable value of the transformational change actions, it should be related to the society needs (e.g., social innovation, intelligent society, Europe and Global society of knowledge, etc.). Thus, the strategic management for a focused technological organization (such as, NIRD, Timişoara) can be thought and proposed to be implemented in the cubic paradigm. The management "in eight corners," or "in eight strategic states/stages" may assure the necessary link between the short-term characteristics to the modern avant-garde managerial approaches (Drucker, 1954), eventually turned into the postmodernist perspective (Kotter, 2012; Putz, 2017), as presented and described in Table 11.2 (see also Chapters 2–4, and 8 of the present monograph).

These change points/states/phases are exemplified on the changing needs in scientific-technological management for the NIRD, Timişoara, indicating, for each stage, their specific content, with the actions designed to assure the successful ranging of the stage. The process as a whole is essential to be followed by the Scientific Director, since he represents the inner changing agent, designated by the General Director, being nevertheless responsible for the management of this complex process, organizationally, scientifically changing (Fig. 11.2).

11.3.1 ADOPTED STRATEGIC MANAGEMENT MODEL

Firstly, the achievement of the *dual* model (from *hierarchy to network*) by the strategic management requires the reorganization of the macromanagement (*bonds* and *relationships*) and the strategic micromanagement (*nodes* and *units*). Then, the competitive, sustainable, and regenerative (CSR) advantage may be acquired in 3D cubic strategic space, in eight management steps within 4 years. It may be done by accelerating the research, development, and innovation (RDI) efficiency from Drucker (1954) to Kotter (2012) toward the strategic cube of the synergic CSR advantage (Putz, 2017, 2019).

11.3.2 TEAM MANAGEMENT

The major Superior Management ways within the governing/boarding team are: *Consultative management* (communication within the governing

TABLE 11.2 The Eight States/Stages in the Strategic Management Cube with Applicability on NIRD, Timişoara.

Stage	Change management, namely, Kottler (2012)	Socio-economic management, namely, Drucker (1954)	Cubic management (Putz, 2017), namely, decision argumentation	Sustainable management, namely, (Bower and Paine) 2017
I	Emergency metaphor	Market standing	*"MARKET TESTING"*	*The manager, as a leader, gives effectiveness to the organization*
II	The business team, the team business	Productivity	*THE BLUE OCEAN*	*The organization can only thrive through its adaptive environmental command*
III	Incentive initiative, Initiative incentive	Profitability	*THE RED OCEAN*	*The organizations become multi-utility in society*
IV	Change volunteers selection on a niche market	Worker performance and attitude	*FEED-BACK*	*The organizations have diverse aims and multiple strategies for them*
V	Removing cultural barriers, accepting profits diminishing and shares declining	Public and social responsibility	*VALUE SHARING*	*Sustainable organizations create value for stake holders*
VI	Teams and finance reorientation for a new investment	Physical and financial resources	*BUSINESS RELAUNCHING MARKETING*	*Multi-stake-holders interactions (shareholders and society) fixe the ethical standards*
VII	Investment acceleration through supply diversification	Innovation	*INCREASING TECHNOLOGICAL QUALITY*	***PESTEcJ(C)*** *integration* *(politic-economic-social-technical-ecological and juridical, possibly also cultural) is essential for sustainability*
VIII	Institutional change through internalization of new production structures, of the relaunched business type	Managerial performance and development	*SECOND CURVE ENTRY. NETWORK LOGISTIC*	*The organization interests toward the distinctive advantage rather than the competitive interest of the shareholders and stakeholders*

Source: Adapted with permission from Putz (2017).

team), *participative management* (working meetings in the governing team), and *delegative management* (by delegating the governing team decisions, on the proposals of the Scientific Director, to the organizational network for changing the hierarchical organization to the dual to cubic management implementation).

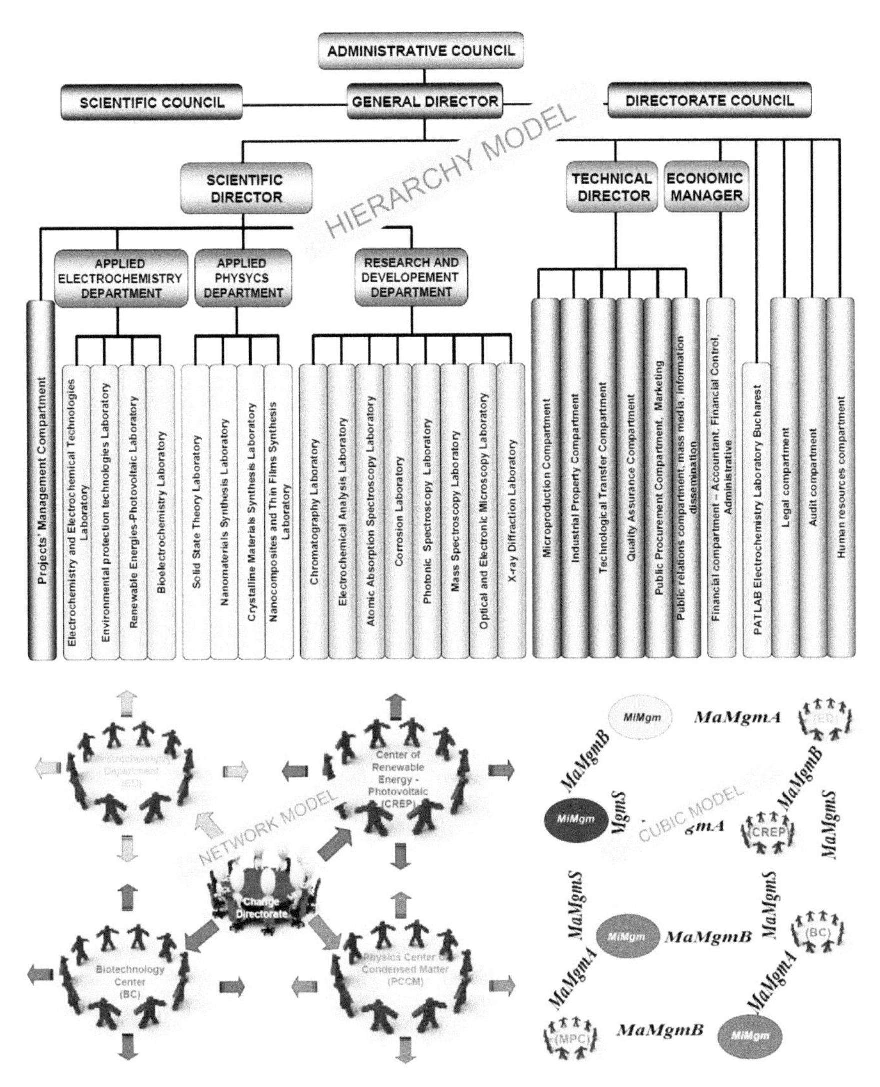

FIGURE 11.2 (See color insert.) Top: hierarchical structure, after Drucker (1954); bottom-left: dual model network structure, after Kotter (2012); bottom-right: 3D model

Cubic Strategic structure, after Putz (2017)—as the present inclusive scientific change paradigm.

11.4 OPERATIONAL AND STRATEGIC OBJECTIVES

I. Of scientific *macromanagement* (*edges in the organizational changing network*):

– *on short term (annual),* ***MaMgmA***:

1. monitoring the *thematic of scientific portfolio/R&D projects* accomplished, from the national plans/programs of research and development, the acquired direct contracts with third parties, the prospected international collaborations
2. valorization of *intellectual property*, along developing the technological transfer chain, either by involvement in spin-offs and start-ups initiatives, or by enhancing the microproduction and services, while finding alternative ways for scientific capitalization (by amplification and acceleration of RDI results)
3. *clustering and hubbing* by patents, ISI (International Scientific Indexing) articles, conferences, and exhibitions, scientific events, and so on
4. optimizing *the staff structure*, of the average earnings, based on a staff rating system (to be delivered)

– *on medium term (bi-annual),* ***MaMgmB***

1. constructing intra- and extrainstitutional *dynamical map of research* (coopetition)
2. perfecting *the information system* (intra- and extraweb communication)
3. relaunching the *scientific marketing* of NIRD, by launching an international periodic journal, and on a periodic (inter)national conference (with Scientific Director as the editor-in-chief, EiC, respectively)
4. continuous (re)defining of *its own scientific product* (branding) of NIRD through an "applicative green" concept,

operational, specific, and original (from the technical readiness level TRL 4 toward TRL 5, with the Technical Director input)

– *on long term (multiannual and strategic),* ***MaMgmS****:*
  1. *initiating organizational change* (by a diagnostic analysis of the changing potential, of the dynamic capacity and capability, both by identifying the competence of incremental and rupture change)
  2. reducing the change resistance by *organizational change motivation*
  3. mobilizing and accelerating the necessary "new" by *organizational changing projection*
  4. implementation of the *organizational change* (transition and transformation achievement), with the controlling intervention to ensure the eco-nanosustainable values enhancement in a systematical manner in NIRD, Timişoara.

II. Of scientific *micro-management* (*vertices/laboratories and projects from the organizational changing network, coupled in order to define the eco-nanosustainable directions for NIRD, Timişoara,* ***MiMgm***:

1. transforming the laboratory into center of *renewable energies—Photovoltaic* (LREP→CREP)
2. testing the photovoltaics with *Fullerenic Graphene* (Nobel Prize in Physics in 2010 for graphene and Nobel Prize in Chemistry in 1996 for fullerene)
3. enhancing the *electrochemistry department* activities
4. initiation of electrochemistry with *molecular machines* (Nobel Prize in Chemistry 2016)
5. transforming the department into physics center of *condensed mater physics* (PDCM→PCCM), with specific renewed scientific research directions
6. initiation of photoelectrochemistry with *optical networks* (Nobel Prize in Physics 2001 for Bose–Einstein Condensate)
7. transforming the *biotechnology* department into center thereof (BD→BC)
8. designing and synthesis of biocompatible sensitizers with *magnetic molecules*.

11.4.1 ACTIONS' SYNERGY

The LREP should be challenged toward transformation while being supported to become Excellence Research Center (CREP), LREP→CREP (Fig. 11.3). This is the *pilot action* in a complex organizational changing environment (with multiple factors/stakeholders) and dynamics (with temporal variable factors) of laboratories and departments of NIRD, Timişoara, on a superior level of research–development–innovation. The corresponding societal recognition, from the cooperation ("all for one and one for all") of the laboratories, departments, to the NIRD, Timişoara directors, having as the inclusive change agent, in the person of the Scientific Director, should be also balanced by appropriate executor delegations, by the General Director of NIRD, Timişoara.

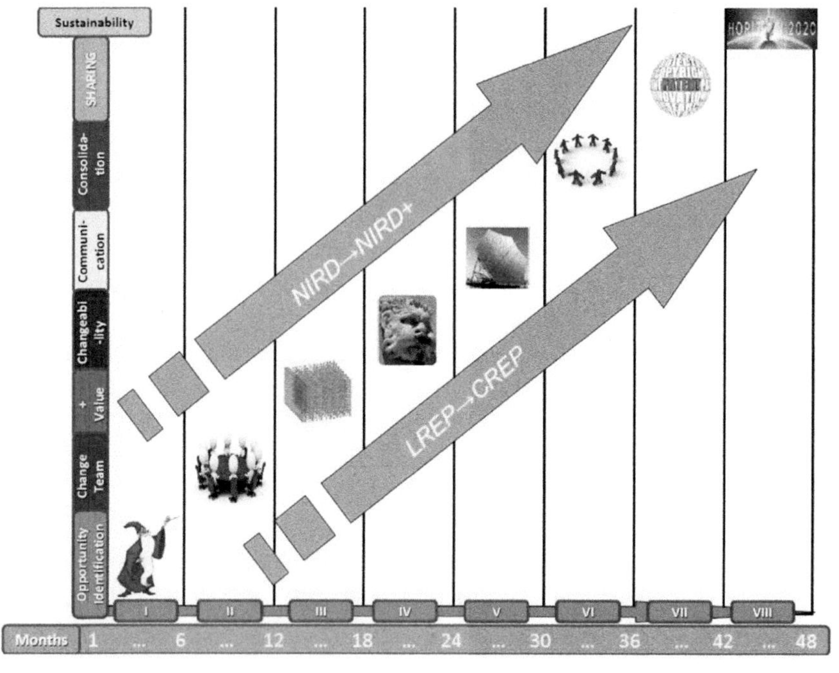

FIGURE 11.3 (See color insert.) Gantt chart of administrative measures implementation identified from Table 11.2 in scientific inclusive change management on NIRD, Timişoara in 4 years (48 months) mandate.

11.5 CONCLUSIONS AND PERSPECTIVES

National organisms of RDI are very close and even the initiators of technological change, due to their very nature and existence. They are the first ones in a society to continually adjust their organization, methodology, personnel in order to include, or even anticipate the new forms of energy and transformation of matter. The new forms of energy and matter will be eventually less consuming for the Natures' resources, driving to renewable, reintegrating, and reusable life cycles. Thus, the management of change methodology and techniques are crucial for their proper functioning. The RDI organizations, like biological organisms can be more than adaptive, so they can be proactive. They shall and, even, prevail in their relation to the economic, industrial, and academic environment toward producing a high social and ecological impact. This behavior brings them prestige, financing (research grants), premium contracts, technological maintenance, productivity, and profitability for their tangible (equipments) values, but even more for its intangible ones—their know-how and the human resources.

Analytically, the change management starting from an immovable, eternal challenging questions, recalling the Kant inquiries to humankind: *What could I know*?; *What should I do*?; *What can I hope for*? Put this way, the change management can offer some strategies of answers, by employing its "cubic" (inclusive) form, since built on 3D, generated, just by these sort of fundamental questions, namely, *What?–How?–When?* They may drive the changing of an organization, or a business toward increasing both efficiency and its efficiency in the specific field of activity or industry. The order of questions is slightly, yet essential, different than the usual one in some literature (Predişcan, 2004); here, the temporal question being preferred at the end, as in relativistic modeling of the continuum space–time, also in accordance with the philosophical order of things (Being and Time/*Sein und Zeit*, namely, Heidegger phenomenology). Under these circumstances, the fundamental axes of the change management, once identified, are "vitalized" by dynamic and progressive answers: *strong* (on the axes), *mixed, or fuzzy* (between axes, i.e., on the plans defined by these ones), and the *synergetic ones* (when joining the origin of all fundamental questions to simultaneous answers on them, on the main diagonal of unitary cube of change). Their extension will fill the space and, by continuous repetition, ensures many cycles of organizational change, in superior inclusive cubes, etc.

This vision is applied to the scientific and technological reality of NIRD, Timişoara, for a continuous development of the executive scientific management, through an inclusive strategic approach, as this cubic one. The agent of the scientific management is, in this case, the scientific manager. For the eight elementary/stages in the strategic cube of organizational change, the measures and methods of monitoring such change have been detailed. Thus, one passes from diagnosis to implementation, while diminishing the changing inertia on workforce, by appropriate personnel training; then one passes to the development of change, paralleling the risk reduction, while increasing the scientific and technical productivity and profitability. Finally, the fulfillment of sustainable strategic objectives of the RDI institute completes the eight-folded (*octavo*) strategic potential of research–development–inclusive innovation.

With this conception of executive scientific management, we answer positively to the degree of flexibility, proper to the changes in a nanotechnologically oriented RDI institute. As a specific illustration, the potential strategic objectives of superior management of NIRD were just unveiled along with the recommendation of necessary adjustments, or acts toward its activation in full competencies and competiveness. The implementation and acceptability of suggested steps of changes, heavily, depends on the organizational feedback (internally), as well as from the external environment (national and international) of academy, economy/industry, and society. There was the continuity in an increasing sustainable spiral, the ability of avoiding crises by anticipation, and by avoiding emergencies by prioritization, closely related to the added value shared by coopetition in an open (scientific and technological) market and future alike.

KEYWORDS

- **axes of the management of change**
- **the change as an opportunity**
- **cubic management**
- **research–development–innovation**
- **hierarchical model of organization**
- **Kotter's dual model of organization**
- **cubic model of organization**

REFERENCES

Bower, J. L.; Paine, L. S. The Error at the Heart of the Organizational Leadership [in Italian as: L'errore al cuore della leadership aziendale] *Harv. Bus. Rev.* (Italia) **2017,** *5*, 24–36. web: http://www.hbritalia.it/maggio-2017/2017/05/02/pdf/gestire-in-unottica-di-lungo-termine-3287/, 2017.

Bruksos, R.; Tumey, P.C. *Turning Change into a Payday: Reinventing Yourself Through the Eight Stages of Change*; Training Consultants: Seattle, 2005.

Drucker, P. F. *The Practice of Management*; Harper Business: New York (Re-issued Ed. 2006), 1954.

Kotler, P. *Marketing Management. Analysis, Planning, Implementation, and Control*, 7th ed.; Prentice-Hall Inc.: New York, 1991.

Kotter, J. P. Accelerate! How the Most Innovative Companies Capitalize on Today's Rapid-Fire Strategic Challenges- and Still Make Their Numbers. *Harv. Bus. Rev.* **2012,** *90* (11), 43–58.

Newton, R. *Managing Change Step By Step: All You Need to Build a Plan and Make it Happen*; Pearson Education Limited: London, 2007.

Predişcan, M. Organizational Change. What, When and How Do We Change? [Originally in Romanian as: *Schimbare Organizaţională. Ce, când şi cum schimbăm?*], West University of Timisoara Publishing House: Timişoara, Romania, 2004.

Putz, M. V. Strategic Cube of the Organization Competitive Advantage. [Originally in Romanian: *Cubul strategic al avantajului competitiv al organizaţiilor*]. MBA Thesis, Faculty of Economy Science and Business Administration, West University of Timişoara, 2017.

Putz, M. V. *Risk Management in Nanotechnology Projects Toward 8th fold Ws*; Chapter 7 of the Present Monograph, 2019.

Index

D

E

F

G

O

P

R

S